Vladimir L. Safonov

Nonequilibrium Magnons

Related Titles

Harry, J. E.

Introduction to Plasma Technology
Science, Engineering and Applications

2011
ISBN: 978-3-527-32763-8

Bhattacharya, R., Paranthaman, M. P. (eds.)

High Temperature Superconductors

2010
ISBN: 978-3-527-40827-6

Radons, G., Rumpf, B., Schuster, H. G. (eds.)

Nonlinear Dynamics of Nanosystems

2010
ISBN: 978-3-527-40791-0

Gubin, S. P. (ed.)

Magnetic Nanoparticles

2009
ISBN: 978-3-527-40790-3

Hofmann, P.

Solid State Physics
An Introduction

2008
ISBN: 978-3-527-40861-0

Moon, F. C.

Applied Dynamics
With Applications to Multibody and Mechatronic Systems

2008
ISBN: 978-3-527-40751-4

Vladimir L. Safonov

Nonequilibrium Magnons

Theory, Experiment, and Applications

WILEY-VCH

WILEY-VCH Verlag GmbH & Co. KGaA

The Author

Dr. Vladimir L. Safonov
Mag & Bio Dynamics, Inc.
1682 Crescent Knolls Gln
Escondido, CA 92029-4200
USA

Library of Congress Card No.:
applied for

British Library Cataloguing-in-Publication Data:
A catalogue record for this book is available from the British Library.

Bibliographic information published by the Deutsche Nationalbibliothek
The Deutsche Nationalbibliothek lists this publication in the Deutsche Nationalbibliografie; detailed bibliographic data are available on the Internet at http://dnb.d-nb.de.

Print ISBN 978-3-527-41117-7

Cover Design Adam-Design, Weinheim
Typesetting le-tex publishing services GmbH, Leipzig
Printing and Binding Markono Print Media Pte Ltd, Singapore

Printed in Singapore

Printed on acid-free paper

to Marina

This book is devoted to the modern theory of magnons, quasiparticles that describe quanta of spin waves in magneto-ordered media.

Contents

Preface

In this book we shall consider magnons, which are quasi particles that describe quanta of spin waves in magneto-ordered media. The concept of magnons as the elementary excitations in ferromagnetic crystals was first introduced by Felix Bloch in 1930. Using this concept, one can represent perturbations of the magnetic material as an evolution of magnon gas excited to some definite level or, classically, as an evolution of spin waves. This theoretical approach helps to explain many thermodynamic, dynamic and kinetic phenomena in magnetic materials.

Magnons are harmonic oscillators in $\boldsymbol{k}$-space. In other words, magnons are good bosons. Recent experimental discovery of Bose–Einstein condensation of quasi equilibrium magnons in YIG film at room temperature directly demonstrated the quantum nature of the elementary magnetic excitations in magneto-ordered systems. The experiments showed that there are conditions at which the macroscopic quantum phenomenon occurs in the situation when just a classical behavior of spin waves was expected.

In this book we use the general language of harmonic motion in application to magnetic dynamic phenomena in magneto-ordered systems. We shall be able to analyze the magnetic dynamics from a more general point of view and find some useful mathematical analogies with other dynamic systems. On the other hand, for those who do not work directly with magnetic systems, this book can provide several ideas and analogies that can be useful in their work and research.

The book contains a large amount of original theoretical results and their comparison with the experimental data. It comprises eight chapters and seven appendices.

Chapter 1 is introductory. It presents the general and universal mathematical language of harmonic oscillators, universal language that unites all Bose-excitation in various physical systems and helps find mathematical analogies between quite different phenomena. Our goal is to briefly discuss basic harmonic oscillator properties and demonstrate simplicity and efficiency of this approach. The hope is that this information will make it possible to read the succeeding chapters without consulting specialized literature.

Basic calculations of phenomenological and microscopic approaches to describe the spin system of a ferromagnet and an antiferromagnet in terms of magnons are given in Chapter 2. We find canonical transformations between the spin deviations and magnon operators of creation and annihilation, and then derive magnon spectra and magnon–magnon interactions.

Chapter 3 is devoted to magnon relaxation analysis. We consider the origin of several mechanisms of magnon relaxation and compare theoretical formulas with the experimental data.

In Chapter 4 the process of parametric excitation of magnon pairs by an external microwave field is described in detail. We take into account specific dynamics of the resonator cavity mode, which plays an important role but has not been considered in earlier studies.

Chapter 5 is devoted to the possibility of a thermodynamic description of a quasi equilibrium magnon system excited parametrically by powerful microwave pulses. The basic idea here is that in a certain rotating coordinate frame the system Hamiltonian becomes stationary and the problem can be reduced to that of a perturbed gas of bosons that approaches thermodynamic equilibrium with a certain effective temperature and chemical potential. We derive and solve basic equations for the resultant states. The thermodynamic approach clearly demonstrates an accumulation of magnons on the bottom of their spectrum which leads to Bose–Einstein condensation of quasi equilibrium magnons.

The phenomenon of Bose–Einstein condensation of quasi equilibrium magnons is analyzed in detail in Chapter 6. We discuss the density of quasi equilibrium magnons and magnon kinetics under the noisy pumping field.

In Chapter 7 we consider an example of the micromagnetic approach in **k**-space for ultrathin ferromagnetic film. We discuss several advantages of this procedure in comparison with the conventional micromagnetic approach in real space.

In Chapter 8 we discuss the collective spin dynamics in nanomagnetic particles. The main focus of this chapter is the use of collective quantum dynamics of clusters of localized electronic spins as a platform for very efficient analog data-processing devices.

Appendix A contains a brief quantum mechanical description of a harmonic oscillator, coherent states, and correspondence between quantum and classical approaches.

Formulas useful for the analytic calculations of dipole–dipole terms in ferromagnetic systems are derived in Appendix B

A general method of constructing nonlinear unitary transformations for a many-body system is described in Appendix C. We show how to eliminate ineffective interaction terms and construct an effective Hamiltonian of interactions.

In Appendix D we derive a dynamic magnetization equation that describes large magnetization motions including magnetization reversal.

In Appendix E we consider an algebra of parametric pairs of magnons and find a single mode realization of this specific state.

Appendix F demonstrates how the so-called modulation method can be used to study instabilities of the stationary state of parametrically excited magnons. There is an effect of amplification of small signals near the threshold of instability.

In Appendix G we consider the experiment demonstrating that incoherent microwaves absorbed by the magnetic system can be transformed to a coherent microwave signal with high efficiency.

This book is written for the readers who have an interest in magnetic dynamics problems both for fundamental and applied research, and for the development of magnetic devices. It is also written in more or less academic style, and therefore can be used by graduate students and professors of technical universities. They can find here topics that may serve as the basis for future doctoral dissertations.

I would like to acknowledge the people who supported my scientific work: Valeri Ozhogin, Andrei Yakubovskii, Hitoshi Yamazaki, Takao Suzuki, Carl Patton, Neal Bertram, and Sakhrat Khizroev.

I have benefited greatly from discussions with my colleagues and friends, Yuri Kalafati, Anatoly Khitrin, Alexander Andrienko and Sergej Demokritov and would like to thank them all.

I want to thank my dear wife and soulmate Marina for her patience, friendship and love. This book is dedicated to her.

Escondido, California, USA, August 2012 *Vladimir L. Safonov*

1
Harmonic Oscillators and the Universal Language of Science

Harmonic motion is one of the most frequently observed phenomena that occurs in nature. We watch oscillations of pendulums, we admire the waves on the tranquil water surface, hear the propagating sound in the air, and see light. All these quite different physical phenomena exhibit local or moving periodic motions of matter described by two trigonometric functions, sine and cosine. This mathematical similarity leads to fundamental concepts of a harmonic oscillator and its "moving brother" the plane wave – a harmonic oscillator in k-space. Both oscillators unify mathematically periodic motions in various physical systems and form a universal language of science [1–5].

Small oscillations of any physical system in the vicinity of equilibrium are harmonic ones as long as the Taylor expansion of the potential energy in this case begins with the quadratic terms [6]. The system of magneto-ordered spins is not an exception. Perturbations of a magnetic order can be regarded as an evolution of waves in a nonlinear medium excited to some definite level or as an evolution of a nonideal gas of quasi particles known as magnons. This approach explains many dynamic, thermodynamic and kinetic phenomena in magnets [7–13].

In this chapter we shall briefly present the basic properties of a harmonic oscillator, its direct connection with the rotation, and introduce variables convenient for the description of spin motion. For simplicity, here we shall use the classical description in a way that resembles the quantum mechanical approach. The goal is to refresh the reader's knowledge of harmonic oscillators and become familiar with the notation used in this book.

1.1
Harmonic Oscillator

The harmonic oscillator is described by the following equation for a generalized coordinate q:

$$\frac{d^2 q}{dt^2} + \omega^2 q = 0 \,. \tag{1.1}$$

Nonequilibrium Magnons, First Edition. Vladimir L. Safonov.

Here ω is the frequency of oscillations. The general solution of (1.1) is

$$q(t) = A_h \cos(\omega t - \phi) , \tag{1.2}$$

where A_h is the amplitude and ϕ is the phase defined by the initial conditions

$$\begin{aligned} q(0) &= A_h \cos\phi , \\ \frac{dq(0)}{dt} &= -A_h \omega \sin\phi . \end{aligned} \tag{1.3}$$

For example, we can obtain (1.1) with

$$\omega = \sqrt{\frac{\kappa}{m}} \tag{1.4}$$

for a mass m on a spring (Figure 1.1a) with the force constant κ using Hamilton's equations of motion

$$\begin{aligned} \frac{dx}{dt} &= \{x, \mathcal{H}\} = \frac{\partial \mathcal{H}}{\partial p} = \frac{p}{m} , \\ \frac{dp}{dt} &= \{p, \mathcal{H}\} = -\frac{\partial \mathcal{H}}{\partial x} = -\kappa x , \end{aligned} \tag{1.5}$$

where

$$\{\mathcal{A}, \mathcal{B}\} \equiv \frac{\partial \mathcal{A}}{\partial x}\frac{\partial \mathcal{B}}{\partial p} - \frac{\partial \mathcal{B}}{\partial x}\frac{\partial \mathcal{A}}{\partial p} \tag{1.6}$$

are the Poisson brackets and the energy of the system, the Hamiltonian, is defined by

$$\mathcal{H} = \frac{p^2}{2m} + \frac{\kappa x^2}{2} . \tag{1.7}$$

Here $q = x$ and $p = mx/dt$ denotes the momentum.

Problem 1.1. Write analogous equations for a charge Q in the LC circuit (Figure 1.1b) denoting $q = Q$, $p = LdQ/dt$, $\kappa = 1/C$ and $\omega = \sqrt{1/LC}$. The reader can continue a number of examples.

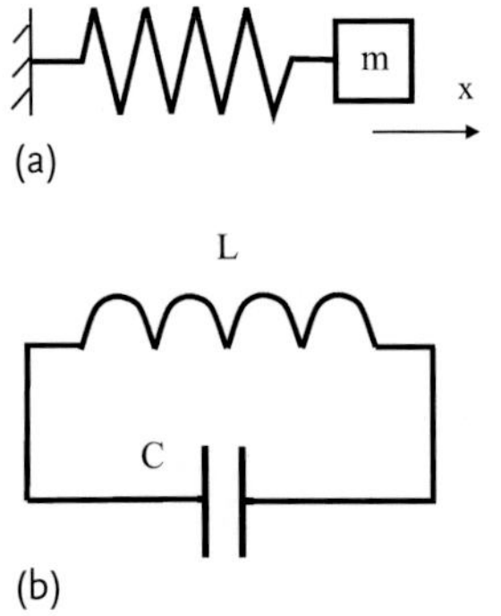

Figure 1.1 (a) Mass on a spring, and (b) LC circuit.

1.1.1 Complex Canonical Variables

Introducing the complex variables,

$$a^* = \frac{1}{2}\left(\frac{x}{x_0} - i\frac{p}{p_0}\right) ,$$
$$a = \frac{1}{2}\left(\frac{x}{x_0} + i\frac{p}{p_0}\right) , \qquad (1.8)$$

where

$$x_0 \equiv \sqrt{\frac{\hbar}{2m\omega}} , \quad p_0 \equiv \sqrt{\frac{\hbar m\omega}{2}} , \qquad (1.9)$$

we obtain classical analogs of creation and annihilation operators that are usually used in quantum mechanics to describe the harmonic motion of a mass on a spring. Here Planck's constant $\hbar$ is used as a dimensional constant in order to have dimensionless a^* and a. The Hamiltonian (1.7) acquires the form

$$\mathcal{H} = \hbar\omega a^* a \qquad (1.10)$$

and the equations in (1.5) become

$$i\frac{da}{dt} = \left[a, \frac{\mathcal{H}}{\hbar}\right]_c = \frac{\partial\mathcal{H}/\hbar}{\partial a^*} = \omega a ,$$
$$i\frac{da^*}{dt} = \left[a^*, \frac{\mathcal{H}}{\hbar}\right]_c = -\frac{\partial\mathcal{H}/\hbar}{\partial a} = -\omega a^* . \qquad (1.11)$$

Here the classical analog of the commutator is defined by

$$[\mathcal{A}, \mathcal{B}]_c \equiv \frac{\partial\mathcal{A}}{\partial a}\frac{\partial\mathcal{B}}{\partial a^*} - \frac{\partial\mathcal{B}}{\partial a}\frac{\partial\mathcal{A}}{\partial a^*} . \qquad (1.12)$$

Note that the equations in (1.11) are another form of the principal equation (1.1) for a harmonic oscillator. As can be seen from the solution of (1.11)

$$a = a(0)\exp(-i\omega t) ,$$
$$a^* = a^*(0)\exp(i\omega t) , \qquad (1.13)$$

complex canonical variables describe rotation (Figure 1.2) and therefore can be convenient for the description of the magnetic moment precession.

Problem 1.2. Write out the complex canonical variables for the LC circuit.

Quantum mechanical properties of a harmonic oscillator are given in Appendix A.

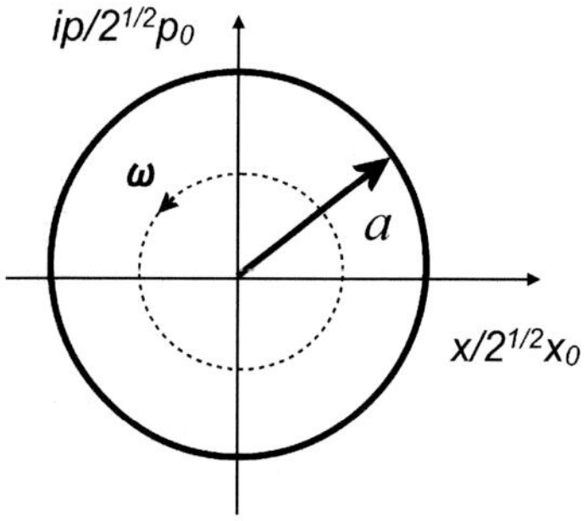

Figure 1.2 Rotation of complex canonical variables.

1.2 Classical Rotation

Hamilton's equations of motion for a particle in classical mechanics can be written as

$$\frac{dx_\ell}{dt} = \{x_\ell, \mathcal{H}\}\,,$$
$$\frac{dp_\ell}{dt} = \{p_\ell, \mathcal{H}\}\,, \tag{1.14}$$

where $\mathcal{H} = \mathcal{H}(x_\ell, p_\ell)$ is the Hamiltonian (energy), $x_\ell = x, y, z$ are coordinates of the particle and $p_\ell = m dx_\ell/dt$ is their momentum (m is the mass).

$$\{\mathcal{A}, \mathcal{B}\} \equiv \sum_\ell \left(\frac{\partial \mathcal{A}}{\partial x_\ell}\frac{\partial \mathcal{B}}{\partial p_\ell} - \frac{\partial \mathcal{B}}{\partial x_\ell}\frac{\partial \mathcal{A}}{\partial p_\ell} \right) \tag{1.15}$$

is the Poisson bracket.

Equations (1.14) can be rewritten in the form which completely resembles quantum Heisenberg equations:

$$i\hbar\frac{dx_\ell}{dt} = [x_\ell, \mathcal{H}]_c\,,$$
$$i\hbar\frac{dp_\ell}{dt} = [p_\ell, \mathcal{H}]_c\,, \tag{1.16}$$

where

$$[\mathcal{A}, \mathcal{B}]_c \equiv i\hbar\{\mathcal{A}, \mathcal{B}\} \tag{1.17}$$

is the classical analog of the commutator. Planck's constant $\hbar$ is introduced as a dimensional constant for simple correspondence with quantum equations. For example, the coordinate and momentum classical commutator corresponds to the quantum commutation rule:

$$[x_\imath, p_\jmath]_c = i\hbar\delta_{\imath\jmath}\,. \tag{1.18}$$

A rotation of particles in a rigid body is associated with their sum of angular momenta (e.g., [6]) $\boldsymbol{J} = \sum \boldsymbol{r} \times \boldsymbol{p}$ or, in the explicit form,

$$J^x = \sum y p_z - z p_y , \quad J^y = \sum z p_x - x p_z , \quad J^z = \sum x p_y - y p_x . \tag{1.19}$$

Taking into account (1.19) and (1.18), we can obtain

$$\begin{aligned} [J^x, J^y]_c &= i\hbar J^z , \\ [J^y, J^z]_c &= i\hbar J^x , \\ [J^z, J^x]_c &= i\hbar J^y . \end{aligned} \tag{1.20}$$

Hamilton's equations for the angular momentum can be written in the following form:

$$i\hbar \frac{d\boldsymbol{J}}{dt} = [\boldsymbol{J}, \mathcal{H}]_c . \tag{1.21}$$

So far as the absolute value $J = |\boldsymbol{J}|$ is conserved, we can introduce new canonical variables J^z and the angle ϕ between the projection of the angular momentum in the x–y plane and the x-axis:

$$\begin{aligned} J^x &= \sqrt{J^2 - (J^z)^2} \cos\phi = \frac{\partial J^y}{\partial \phi} , \\ J^y &= \sqrt{J^2 - (J^z)^2} \sin\phi = -\frac{\partial J^x}{\partial \phi} . \end{aligned} \tag{1.22}$$

It is easy to check that the Poisson bracket in this case becomes

$$\{\mathcal{A}, \mathcal{B}\} \equiv \frac{\partial \mathcal{A}}{\partial \phi}\frac{\partial \mathcal{B}}{\partial J^z} - \frac{\partial \mathcal{B}}{\partial \phi}\frac{\partial \mathcal{A}}{\partial J^z} . \tag{1.23}$$

If the energy depends on the J_z component only, there is a gyroscopic precession around the z-axis. The equations of this precession follow from (1.21) and (1.23):

$$\frac{dJ_x}{dt} = -J_y \frac{\partial \mathcal{H}}{\partial J_z} , \quad \frac{dJ_y}{dt} = J_x \frac{\partial \mathcal{H}}{\partial J_z} . \tag{1.24}$$

In vector form we have

$$\frac{d\boldsymbol{J}}{dt} = \boldsymbol{J} \times \left(-\frac{\partial \mathcal{H}}{\partial \boldsymbol{J}} \right) . \tag{1.25}$$

1.2.1
Classical Spin and Magnetic Resonance

Now we shall consider a classical spin momentum $\hbar \boldsymbol{S}$ which has properties analogous to the angular momentum $\boldsymbol{J}$. Although spin is a purely quantum mechanical concept with no true analog in classical physics, an introduction of "classical spin" $\boldsymbol{S}$ helps to understand the structure of equations and some details of spin dynamics. In the next chapter the classical spins will be associated with the local magnetic moments of the magneto-ordered material.

The dynamic equation for the dimensionless spin variable has the form

$$i\frac{d}{dt}\boldsymbol{S} = \left[\boldsymbol{S}, \frac{\mathcal{H}}{\hbar}\right]_c , \tag{1.26}$$

with the following spin commutation rules:

$$\begin{aligned} [S^x, S^y]_c &= i S^z , \\ [S^y, S^z]_c &= i S^x , \\ [S^z, S^x]_c &= i S^y . \end{aligned} \tag{1.27}$$

Classically, the magnetic dipole moment $\boldsymbol{\mu}$ appears as a result of the rotation of charged particle. It equals a constant times the angular momentum

$$\boldsymbol{\mu} = -\hbar\gamma \boldsymbol{S} , \tag{1.28}$$

where γ is the so-called gyromagnetic ratio, and the negative sign corresponds to electronic spins. Thus the energy of spin in the magnetic field $\boldsymbol{H} = (0, 0, H)$ is equal to

$$\mathcal{H} = \frac{(\hbar \boldsymbol{S})^2}{2 I_S} + \hbar\gamma H S^z , \tag{1.29}$$

where the first term describes the kinetic energy (I_S denotes the moment of inertia) and the second term describes the interaction of magnetic dipole moment with the magnetic field $-(\boldsymbol{\mu} \cdot \boldsymbol{H})$. Taking into account that S = const., the kinetic energy $\propto \boldsymbol{S}^2$ is the integral of motion and may always be omitted.

Introducing the circular spin components $S^{\pm} = S^x \pm i S^y$, from (1.27) we get

$$\begin{aligned} [S^z, S^{\pm}]_c &= \pm S^{\pm} , \\ [S^+, S^-]_c &= 2S^z . \end{aligned} \tag{1.30}$$

From the equation of motion (1.26) and (1.29) follows

$$i\frac{d}{dt}S^- = \left[S^-, \frac{\mathcal{H}}{\hbar}\right]_c = -\omega S^- \tag{1.31}$$

and

$$i\frac{d}{dt}S^+ = \left[S^+, \frac{\mathcal{H}}{\hbar}\right]_c = \omega S^+ . \tag{1.32}$$

Their solution

$$\begin{aligned} S^- &= S^-(0)\exp(-i\omega t) , \\ S^+ &= S^+(0)\exp(i\omega t) , \end{aligned} \tag{1.33}$$

describes the spin rotation with the frequency of magnetic resonance (Figure 1.3)

$$\omega = \gamma H . \tag{1.34}$$

So far as

$$[S^z, \mathcal{H}]_c = 0 ,$$

this magnetic resonance can be excited by the alternating field applied in the x–y plane only.

The solution (1.33) for the circular spin components is similar to the solution for the complex variables (1.13). This means that the spin motion can be represented in terms of harmonic oscillator variables. Holstein and Primakoff [8] were the first who understood the convenience of this representation and found it in the form:

$$\begin{aligned} S^z &= -S + a^* a , \\ S^+ &= a^* \sqrt{2S - a^* a} , \\ S^- &= \sqrt{2S - a^* a} a . \end{aligned} \tag{1.35}$$

It is easy to check that all spin commutations (1.27) and (1.30) are valid for the commutator (1.12) expressed in terms of complex variables. The Hamiltonian (1.29) can be rewritten as

$$\mathcal{H} = \text{const} + \hbar\omega a^* a . \tag{1.36}$$

If we change the direction of magnetic field $H \to -H$, the transformation (1.35) becomes

$$\begin{aligned} S^z &= S - a^* a , \\ S^+ &= \sqrt{2S - a^* a} a , \\ S^- &= a^* \sqrt{2S - a^* a} . \end{aligned} \tag{1.37}$$

It is also convenient to represent the equation of motion for the spin in the form of the Landau–Lifshitz equation [14] without damping:

$$\frac{d}{dt}\mathbf{S} = \mathbf{S} \times \gamma \mathbf{H}_{\text{eff}} , \tag{1.38}$$

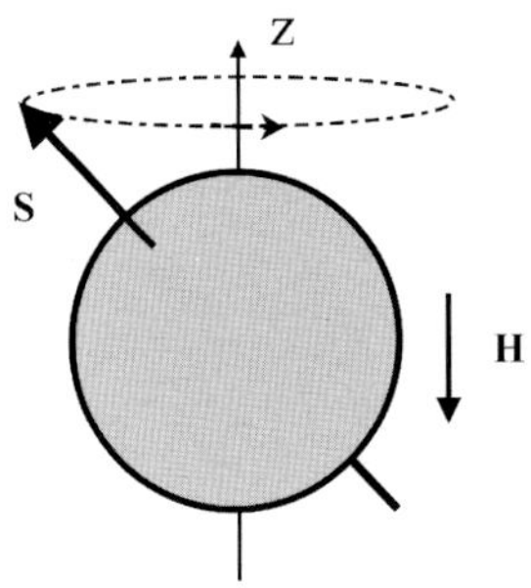

Figure 1.3 Magnetic resonance. Precession of spin in a magnetic field.

where

$$\gamma \boldsymbol{H}_{\text{eff}} = -\frac{\partial \mathcal{H}/\hbar}{\partial \boldsymbol{S}}$$

is the effective magnetic field. The absolute value of spin is conserved in (1.38) and the spin precesses around the effective field.

The principal difference between the angular momentum and spin appears when energy losses are taken into account. Friction in classical rotation leads to a decrease of both angular momentum and the rotation frequency until the system comes to a complete stop. At the same time, the absolute value of spin is conserved during the damping of spin motion with the same frequency of magnetic resonance.

1.3
Collective Variables and Harmonic Oscillators in *k*-space

Now we move to a harmonic oscillator in *k*-space which describes plane waves, another version of harmonic motion in nature. The equation of this oscillator has the form

$$\frac{d^2 q_k}{dt^2} + \omega_k^2 q_k = 0 \,, \tag{1.39}$$

and it differs from (1.1) by the presence of k, the wave-vector index for the generalized coordinate q_k and frequency ω_k.

1.3.1
Chain of Masses and Springs

Let us consider a mechanical system, the infinite chain containing point masses m coupled by springs (Figure 1.4a) with the same force constant κ and length d_c. This chain represents a simple model of a one-dimensional crystal, in which $x_\ell = d_c \cdot \ell$ is the equilibrium position of the ℓth mass and u_ℓ is its displacement. The equation of motion has the form:

$$m \frac{d^2 u_\ell}{dt^2} = -\kappa(2u_\ell - u_{\ell-1} - u_{\ell+1}) \,. \tag{1.40}$$

Introducing the collective variable U_k as

$$u_\ell = U_k \exp(ikx_\ell) \,, \tag{1.41}$$

where k is the wave vector, we obtain

$$\begin{aligned} m \frac{d^2 U_k}{dt^2} &= -\kappa(2 - e^{ikd_c} - e^{-ikd_c}) U_k \\ &= -4\kappa \sin^2\left(\frac{kd_c}{2}\right) U_k \,. \end{aligned} \tag{1.42}$$

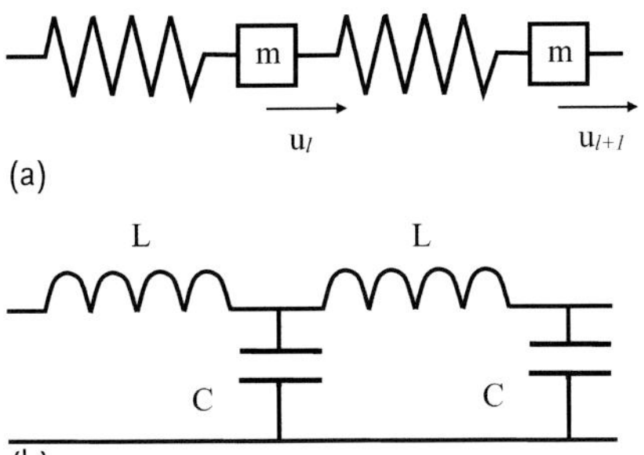

Figure 1.4 (a) Cell of the mass on a spring chain, and (b) cell of the LC chain.

Thus, the collective motion of the chain can be expressed in terms of harmonic oscillator (1.39) with $q_k = U_k$ and the frequency

$$\omega_k = 2\left|\sin\left(\frac{k d_c}{2}\right)\right|\sqrt{\frac{\kappa}{m}}\,. \tag{1.43}$$

In other words, we have obtained the spectrum ω_k of elastic waves in the chain. Note that

$$0 \le \omega_k \le \omega_b\,, \qquad \omega_b \equiv 2\sqrt{\frac{\kappa}{m}}\,. \tag{1.44}$$

This indicates a band structure of excitations in the lattice system.

One can write analogous equations for the LC circuit chain (Figure 1.4b) and obtain

$$\omega_k = 2\left|\sin\left(\frac{k d_c}{2}\right)\right|\sqrt{\frac{1}{LC}}\,. \tag{1.45}$$

Problem 1.3. Write out the complex variables a_k^* and a_k for the (i) chain of masses and springs, (ii) chain of LC circuits.

1.3.2
Chain of Magnetic Particles

Consider now a chain of magnetic particles along the x-axis (Figure 1.5) with coordinates $x_\ell = d_c \cdot \ell$. Each particle represents a classical spin $\boldsymbol{S}$ oriented along the z-axis at the equilibrium. We shall write the energy of the chain in the form

$$\mathcal{H} = -\frac{\hbar\gamma H_K S}{2}\sum_j\left(\frac{S_j^z}{S}\right)^2 - \frac{J}{2}\sum_j\left(\boldsymbol{S}_j\cdot\boldsymbol{S}_{j-1}\right)\,. \tag{1.46}$$

Here the first term describes the uniaxial anisotropy with the field H_K along the z-axis and the second term corresponds to the exchange interaction between nearest

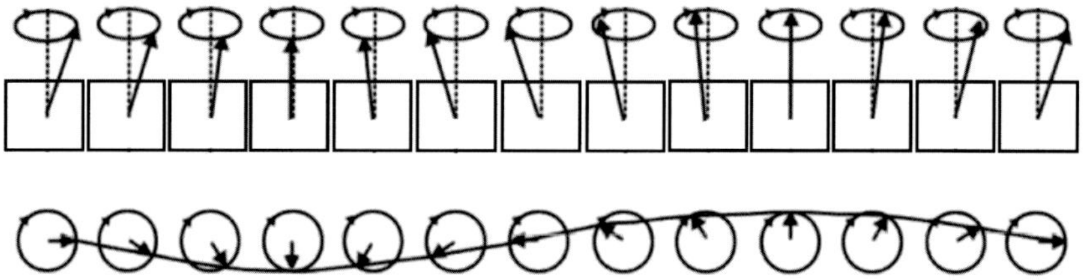

Figure 1.5 Spin wave in the chain of magnetic particles.

neighbors with the exchange integral J. In order to represent $\mathcal{H}$ in terms of harmonic oscillations, we shall use the transformation (1.37) in the approximate form

$$\begin{aligned} S_j^z &= S - a_j^* a_j \,, \\ S_j^+ &\simeq \sqrt{2S}\, a_j \,, \\ S_j^- &\simeq \sqrt{2S}\, a_j^* \,. \end{aligned} \tag{1.47}$$

Taking into account that

$$(\boldsymbol{S}_j \cdot \boldsymbol{S}_{j-1}) = S_j^z S_{j-1}^z + \frac{1}{2}\left(S_j^+ S_{j-1}^- + S_j^- S_{j-1}^+\right) \,, \tag{1.48}$$

we can find the quadratic term of the complex variables part of (1.46):

$$\begin{aligned} \mathcal{H}^{(2)} &= \hbar\gamma H_\mathrm{K} \sum_j a_j^* a_j \\ &\quad + \frac{JS}{2}\sum_j \left(a_j^* a_j + a_{j-1}^* a_{j-1} - a_j^* a_{j-1} - a_{j-1}^* a_j\right) \,. \end{aligned} \tag{1.49}$$

The dynamic equation for the complex variable in this case is

$$\begin{aligned} i\frac{d a_j}{dt} &= \frac{\partial \mathcal{H}^{(2)}/\hbar}{\partial a_j^*} \\ &= \gamma H_\mathrm{K} a_j + \frac{JS}{2}(2a_j - a_{j-1} - a_{j+1}) \,. \end{aligned} \tag{1.50}$$

Introducing the collective variable

$$a_j = a_k \exp(i k x_j) \,, \tag{1.51}$$

we simplify (1.50) to the form of a harmonic oscillator in k-space

$$i\frac{d a_k}{dt} = \omega_k a_k \,, \tag{1.52}$$

where

$$\begin{aligned} \omega_k &= \gamma H_\mathrm{K} + \frac{JS}{2}(2 - e^{ikd_\mathrm{c}} - e^{-ikd_\mathrm{c}}) \\ &= \gamma H_\mathrm{K} + 2JS \sin^2\left(\frac{k d_\mathrm{c}}{2}\right) \end{aligned} \tag{1.53}$$

is the frequency of collective spin excitation – spin wave with wave vector k.

Once we have transformed the spin components in (1.46) to the complex variables (analog of creation and annihilation operators), we are ready to work with a system of harmonic oscillators:

$$\mathcal{H} = \sum_k \hbar\omega_k a_k^* a_k + \text{nonlinear terms.} \tag{1.54}$$

Appendix A provides additional information on the quantum mechanics of a harmonic oscillator.

Problem 1.4. Find the magnon spectrum in the chain of magnetic particles (1.46) for the case when the magnetic field $\boldsymbol{H}$ is applied along the z-axis.

Problem 1.5. Find the magnon spectrum in the chain of magnetic particles with the uniaxial anisotropy oriented along the chain.

1.4 Discussion

a) Why are the complex canonical variables so useful?
b) Is it possible to decouple two oscillators $\hbar\omega_1 a_1^* a_1$ and $\hbar\omega_2 a_2^* a_2$ with the interaction term $V(a_1^* a_2 + a_2^* a_1)$?
c) What are the similarities and differences between the angular momentum and spin?
d) Which solution is preferred (and why) for computer simulations of the magnetic dynamics in the chain of magnetic particles: (i) solution of dynamic equations for individual particles interacting with their neighbors or, (ii) solution of magnon (collective) dynamics and then application of this solution to the local magnetic moment?

2
Magnons in Ferromagnets and Antiferromagnets

Spin waves represent a set of oscillators (normal modes) with frequencies ω_k in $\boldsymbol{k}$-space that describe transverse deviations of magnetization from the magnetically ordered "vacuum" state. By convention, the quantum of spin wave is a quasi particle, which is called a magnon. A magnon's energy is equal to $\hbar\omega_k$, where $\hbar$ is Planck's constant. According to quantum mechanics, the magnetic moment of the system is equal to the negative partial derivative of the energy in the magnetic field [15, 16]. Thus, a magnon carries the magnetic moment

$$\mu_k = -\frac{\partial \hbar \omega_k}{\partial H}\,, \tag{2.1}$$

where $\boldsymbol{H}$ is the external magnetic field. Perturbations of the magnetic system can be regarded as an evolution of a magnon gas and this approach helps to explain many thermodynamic, dynamic, and kinetic phenomena in magnetics. In this chapter we shall consider examples of calculations of magnon spectra in the framework of phenomenological theory and microscopic modeling.

As it follows from the microscopic theory (see, for example, [9, 12, 17–19]), exchange interaction between the nearest localized spins in the lattice is the origin of magnetic order in crystals. The sign of the exchange integral provides two principally different mutual orientations of the nearest spins: parallel (ferromagnetic order) and antiparallel (antiferromagnetic order), and the magnitude of the exchange integral is responsible for the temperature of the phase transition. There are many different solids in nature which exhibit various magnetic order: ferromagnets, antiferromagnets, and ferrites.

The magnetic order of the simplest magnetic system, the ferromagnet, is characterized by parallel orientation of all spins. Complex magnetic structures usually are modeled as combinations of interacting ferromagnetic sublattices inserted into each other [20]. It is possible also to develop an approach which does not use a model representation [21].

Nonequilibrium Magnons, First Edition. Vladimir L. Safonov.

2.1 Phenomenological Description

Any phenomenological approach includes two important parts: (i) a choice of principal variables, their derivatives, and the basic physical equations for these variables, and (ii) a construction of the most general function of state (like energy or Lagrangian) using general symmetry considerations [20, 22].

The ferromagnetic energy density is represented as a sum of various combinations of the magnetization $\boldsymbol{M}(\boldsymbol{r})$ and its spatial derivatives $\partial M_j/\partial x_\ell$ with unknown coefficients. The symmetry operations that are allowed in the system should not change the energy representation and this condition excludes unnecessary terms.

The inversion of time symmetry $t \to -t$ changes the sign of $\boldsymbol{M} \to -\boldsymbol{M}$ (analogous to the angular momentum $\boldsymbol{J} \to -\boldsymbol{J}$). Thus, the energy must remain even with respect to M_j terms:

$$a_1 M_x^2 + a_2 M_y^2 + a_3 M_z^2 + a_4 \left(\frac{\partial M_x}{\partial x}\right)^2 + a_5 \frac{\partial M_x}{\partial x}\frac{\partial M_y}{\partial x} + a_6 \frac{\partial M_x}{\partial y}\frac{\partial M_y}{\partial x} + \ldots$$

As an example let us consider a 90° rotation symmetry around the z-axis. In this case we have the following transformation:

$$x \to y\,, \quad y \to -x\,, \quad z \to z$$
$$M_x \to M_y\,, \quad M_y \to -M_x\,, \quad M_z \to M_z$$

which brings the change of sign for

$$\frac{\partial M_x}{\partial x}\frac{\partial M_y}{\partial x} + \frac{\partial M_x}{\partial y}\frac{\partial M_y}{\partial y} \to -\frac{\partial M_y}{\partial y}\frac{\partial M_x}{\partial y} - \frac{\partial M_y}{\partial x}\frac{\partial M_x}{\partial x}\,.$$

This is obviously possible if the coefficient in front of this combination is equal to zero.

Applying other symmetry transformations, one can obtain the magnetic density form corresponding to a given crystal. This energy contains important information on magnon spectra and magnon–magnon interactions.

2.1.1 Magnons in a Ferromagnet

We shall write the magnetic energy (Hamiltonian) of the ferromagnet in the following form:

$$\mathcal{H} = \int (\mathcal{U}_{\text{ex}} + \mathcal{U}_{\text{anis}} + \mathcal{U}_{\text{Z}} + \mathcal{U}_{\text{dmag}})\, d\boldsymbol{r}\,. \tag{2.2}$$

Here

$$\mathcal{U}_{\text{ex}} = \frac{A_{\text{ex}}}{M_{\text{s}}^2}\left[\left(\frac{\partial \boldsymbol{M}}{\partial x}\right)^2 + \left(\frac{\partial \boldsymbol{M}}{\partial y}\right)^2 + \left(\frac{\partial \boldsymbol{M}}{\partial z}\right)^2\right] \tag{2.3}$$

is the exchange energy density for cubic symmetry,

$$\mathcal{U}_{\text{anis}} = -K_{\text{u}} \left(\frac{M_z}{M_{\text{s}}} \right)^2 \tag{2.4}$$

is the uniaxial anisotropy energy density,

$$\mathcal{U}_{\text{Z}} = -\boldsymbol{H} \cdot \boldsymbol{M} \tag{2.5}$$

is the Zeeman energy density, and $\mathcal{U}_{\text{dmag}}$ is the magnetostatic energy density.

We shall consider a continuous ferromagnetic sample as a system of coupled cubic grains of the volume $V = d_{\text{c}}^3$, where d_{c} is the linear size of each cube (cell). Each cube can be characterized as a classical spin:

$$\boldsymbol{S}_j \equiv -\frac{\boldsymbol{M}_j V}{\hbar\gamma} , \tag{2.6}$$

where $\boldsymbol{M}_j$ is the magnetization vector of the jth cube and $\gamma > 0$ is the gyromagnetic ratio. $|\boldsymbol{M}_j| = M_{\text{s}}$, where M_{s} is the saturation magnetization, and therefore, $|\boldsymbol{S}_j| = S \equiv M_{\text{s}} V/\hbar\gamma$. Each spin is localized in the center of the cube (Figure 2.1). The characterization of magnetization in terms of an effective spin is classical. However the spin notation is convenient for numerical simulations and it also provides a bridge to quantum spin models.

The Hamiltonian (2.2) of the system in terms of spin notation can be written as:

$$\mathcal{H} = \mathcal{H}_{\text{ex}} + \mathcal{H}_{\text{anis}} + \mathcal{H}_{\text{Z}} + \mathcal{H}_{\text{dmag}} . \tag{2.7}$$

Here

$$\mathcal{H}_{\text{ex}} = -\frac{J}{2} \sum_{j \neq j'}^{(\text{n.n.})} (\boldsymbol{S}_j \cdot \boldsymbol{S}_{j'}) \tag{2.8}$$

describes the exchange interaction between nearest neighbors, $J = 2A_{\text{ex}} d_{\text{c}}/S^2$. Note that this form is identical to the usual micromagnetic approximation that assumes a linear variation of the magnetization between the grain centers.

$$\mathcal{H}_{\text{anis}} = -\frac{\hbar\gamma\, H_{\text{K}} S}{2} \sum_{j=1}^{\mathcal{N}} \left(\frac{S_j^z}{S} \right)^2 \tag{2.9}$$

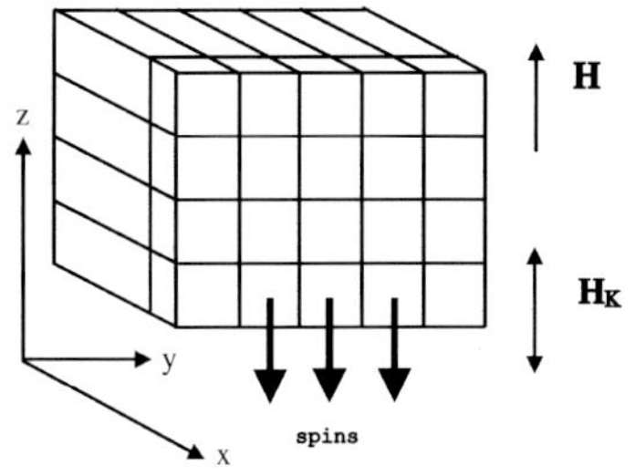

Figure 2.1 Ferromagnetic sample.

is the uniaxial anisotropy energy, where $H_K = 2K_u/M_s$ is the anisotropy field and $\mathcal{N}$ is the number of elementary cells (cubes) in the sample.

$$\mathcal{H}_Z = \hbar\gamma \left(\boldsymbol{H} \cdot \sum_{j=1}^{\mathcal{N}} \boldsymbol{S}_j \right) \tag{2.10}$$

is the Zeeman energy.

The magnetostatic interaction can be approximated by the dipole–dipole interaction energy

$$\mathcal{H}_{\text{dmag}} = \frac{(\hbar\gamma)^2}{2} \sum_{j \neq j'} \left[\frac{(\boldsymbol{S}_j \cdot \boldsymbol{S}_{j'})}{r_{jj'}^3} - \frac{3(\boldsymbol{S}_j \cdot \boldsymbol{r}_{jj'})(\boldsymbol{S}_{j'} \cdot \boldsymbol{r}_{jj'})}{r_{jj'}^5} \right] \tag{2.11}$$

where $\boldsymbol{r}_{jj'} = \boldsymbol{r}_j - \boldsymbol{r}_{j'}$ and $\boldsymbol{r}_j$, $\boldsymbol{r}_{j'}$ are the radius vectors of the j and j' spins, respectively.

2.1.1.1 Holstein–Primakoff Transformation

Now we shall represent (2.7) as the Hamiltonian of a nonlinear system of oscillators using the Holstein–Primakoff transformation

$$\begin{aligned} S_j^z &= -S + a_j^\dagger a_j \,, \\ S_j^+ &\simeq \sqrt{2S} a_j^\dagger - \frac{a_j^\dagger a_j^\dagger a_j}{\sqrt{8S}} \,, \\ S_j^- &\simeq \sqrt{2S} a_j - \frac{a_j^\dagger a_j a_j}{\sqrt{8S}} \,, \end{aligned} \tag{2.12}$$

where $a_j^\dagger$ and a_j are creation and annihilation Bose operators, the quantum analogs of classical complex variables. Here we use the so-called procedure of canonical quantization to preserve the formal structure of the classical theory and proceed from classical variables $\boldsymbol{S}_j$ to quantum operators $a_j^\dagger$ and a_j.

This transformation makes it possible to rewrite the initial Hamiltonian in the form

$$\mathcal{H} \simeq \mathcal{H}^{(0)} + \mathcal{H}^{(1)} + \mathcal{H}^{(2)} + \mathcal{H}^{(3)} + \mathcal{H}^{(4)} + \ldots \,, \tag{2.13}$$

where the upper number indicates the number of normally ordered Bose operators.

Let us perform a set of calculations:

$$\begin{aligned} S_j^z S_{j'}^z &= \left(-S + a_j^\dagger a_j\right)\left(-S + a_{j'}^\dagger a_{j'}\right) \\ &= S^2 - S\left(a_j^\dagger a_j + a_{j'}^\dagger a_{j'}\right) + a_j^\dagger a_{j'}^\dagger a_j a_{j'} \,, \end{aligned} \tag{2.14}$$

$$\begin{aligned} S_j^+ S_{j'}^- &\simeq \left(\sqrt{2S} a_j^\dagger - \frac{a_j^\dagger a_j^\dagger a_j}{\sqrt{8S}}\right)\left(\sqrt{2S} a_{j'} - \frac{a_{j'}^\dagger a_{j'} a_{j'}}{\sqrt{8S}}\right) \\ &\simeq 2S a_j^\dagger a_{j'} - \frac{1}{2}\left(a_j^\dagger a_j^\dagger a_j a_{j'} + a_j^\dagger a_{j'}^\dagger a_{j'} a_{j'}\right) \,, \end{aligned} \tag{2.15}$$

$$\left(S_j^z\right)^2 = \left(-S + a_j^\dagger a_j\right)^2 = S^2 - 2S a_j^\dagger a_j + a_j^\dagger a_j a_j^\dagger a_j$$
$$= S^2 - (2S-1) a_j^\dagger a_j + a_j^\dagger a_j^\dagger a_j a_j \,. \tag{2.16}$$

The previous equations used the commutation relation $a_j a_{j'}^\dagger = a_{j'}^\dagger a_j + \delta_{jj'}$.

The next group of calculations is

$$\begin{aligned}
&(\boldsymbol{S}_j \cdot \boldsymbol{r}_{jj'}) (\boldsymbol{S}_{j'} \cdot \boldsymbol{r}_{jj'}) \\
&= \left[S_j^z z_{jj'} + \frac{1}{2}\left(S_j^+ r_{jj'}^- + S_j^- r_{jj'}^+\right)\right]\left[S_{j'}^z z_{jj'} + \frac{1}{2}\left(S_{j'}^+ r_{jj'}^- + S_{j'}^- r_{jj'}^+\right)\right] \\
&= (z_{jj'})^2 S_j^z S_{j'}^z + \frac{1}{4} r_{jj'}^+ r_{jj'}^- \left(S_j^+ S_{j'}^- + S_j^- S_{j'}^+\right) \\
&\quad + \frac{1}{2} z_{jj'} r_{jj'}^- \left(S_j^z S_{j'}^+ + S_{j'}^z S_j^+\right) + \frac{1}{2} z_{jj'} r_{jj'}^+ \left(S_j^z S_{j'}^- + S_{j'}^z S_j^-\right) \\
&\quad + \frac{1}{4}\left(r_{jj'}^-\right)^2 S_j^+ S_{j'}^+ + \frac{1}{4}\left(r_{jj'}^+\right)^2 S_j^- S_{j'}^- \,,
\end{aligned} \tag{2.17}$$

where

$$r_{jj'}^\pm = x_{jj'} \pm i y_{jj'} \,, \tag{2.18}$$

$$\begin{aligned}
S_j^z S_{j'}^+ &\simeq -S\sqrt{2S} a_{j'}^\dagger + \sqrt{2S} a_{j'}^\dagger a_j^\dagger a_j + \sqrt{\frac{S}{8}} a_{j'}^\dagger a_{j'}^\dagger a_{j'} \\
S_j^z S_{j'}^- &\simeq -S\sqrt{2S} a_{j'} + \sqrt{2S} a_j^\dagger a_j a_{j'} + \sqrt{\frac{S}{8}} a_{j'}^\dagger a_{j'} a_{j'} \,,
\end{aligned} \tag{2.19}$$

$$\begin{aligned}
S_j^- S_{j'}^- &\simeq 2S a_j a_{j'} - \frac{1}{2}\left(a_{j'}^\dagger a_{j'} a_{j'} a_j + a_j^\dagger a_j a_j a_{j'}\right) \,, \\
S_j^+ S_{j'}^+ &\simeq 2S a_j^\dagger a_{j'}^\dagger - \frac{1}{2}\left(a_j^\dagger a_{j'}^\dagger a_{j'}^\dagger a_{j'} + a_{j'}^\dagger a_j^\dagger a_j^\dagger a_j\right) \,.
\end{aligned} \tag{2.20}$$

Substituting all spin combinations in (2.8)–(2.11) with corresponding combinations of Bose operators (2.14)–(2.20), we can write the explicit forms of $\mathcal{H}^{(n)}$.

For convenience, we also introduce collective coordinates using Fourier transformation

$$\begin{aligned}
a_j &= \frac{1}{\sqrt{\mathcal{N}}} \sum_k a_k \exp(i\boldsymbol{k}\boldsymbol{r}_j) \,, \\
a_j^\dagger &= \frac{1}{\sqrt{\mathcal{N}}} \sum_k a_k^\dagger \exp(-i\boldsymbol{k}\boldsymbol{r}_j) \,.
\end{aligned} \tag{2.21}$$

One can check the following commutation relations

$$\begin{aligned}
[a_k, a_q] &= \left[a_k^\dagger, a_q^\dagger\right] = 0 \,, \\
\left[a_k, a_q^\dagger\right] &= \Delta(\boldsymbol{k} - \boldsymbol{q}) \,,
\end{aligned} \tag{2.22}$$

where $\Delta(\boldsymbol{k})$ is the Kronecker delta function:

$$\Delta(\boldsymbol{k}) = \begin{cases} 1 \text{ if } \boldsymbol{k} = 0 \,, \\ 0 \text{ if } \boldsymbol{k} \neq 0 \,. \end{cases} \tag{2.23}$$

The zeroth ground energy term has the form

$$\mathcal{H}^{(0)} = -\frac{\mathcal{N} S^2}{2}\left[J Z_0 - (\hbar\gamma)^2 \sum_{\boldsymbol{r}}\left(\frac{1}{r^3} - \frac{3z^2}{r^5}\right)\right] - H \mathcal{N} S \hbar\gamma\left(\frac{H_\mathrm{K}}{2} + H\right) , \tag{2.24}$$

where $Z_0 = \sum^{\text{n.n.}} 1 = 6$ is the number of nearest neighbors in the cube lattice.

The first-order term can be written as:

$$\mathcal{H}^{(1)} = \frac{3(\hbar\gamma)^2 S}{4}\sqrt{\frac{2S}{\mathcal{N}}}\sum_{\boldsymbol{k}} \left[\left(\sum_{j\neq j'} \frac{z_{jj'} r^-_{jj'}}{r^5_{jj'}}\left[\exp(i\boldsymbol{k}\boldsymbol{r}_j) + \exp(i\boldsymbol{k}\boldsymbol{r}_{j'})\right]\right) a^\dagger_{\boldsymbol{k}} + \left(\sum_{j\neq j'} \frac{z_{jj'} r^+_{jj'}}{r^5_{jj'}}\left[\exp(-i\boldsymbol{k}\boldsymbol{r}_j) + \exp(-i\boldsymbol{k}\boldsymbol{r}_{j'})\right]\right) a_{\boldsymbol{k}}\right] . \tag{2.25}$$

Taking into account that

$$\sum_j \exp(i\boldsymbol{k}\boldsymbol{r}_j) = \sum_j \exp(-i\boldsymbol{k}\boldsymbol{r}_j) = \mathcal{N}\Delta(\boldsymbol{k}) , \tag{2.26}$$

we obtain

$$\sum_{j\neq j'} \frac{z_{jj'} r^\pm_{jj'}}{r^5_{jj'}} \exp(\mp i\boldsymbol{k}\boldsymbol{r}_j) = \sum_{\boldsymbol{r}} \frac{z r^\pm}{r^5}\mathcal{N}\Delta(\boldsymbol{k}) , \tag{2.27}$$

where $\boldsymbol{r} \equiv \boldsymbol{r}_{jj'}$. As a result, the first-order term becomes

$$\mathcal{H}^{(1)} = \frac{3(\hbar\gamma)^2 S}{2}\sqrt{2S\mathcal{N}}\left[\left(\sum_{\boldsymbol{r}} \frac{z r^-}{r^5}\right) a^\dagger_0 + \left(\sum_{\boldsymbol{r}} \frac{z r^+}{r^5}\right) a_0\right] . \tag{2.28}$$

So far as the z-axis is the symmetry axis, the sums equal zero.

$$\sum_{\boldsymbol{r}} \frac{z r^\pm}{r^5} = \sum_{\boldsymbol{r}} \frac{z(x \pm i y)}{r^5} = 0 .$$

In this case the ground state energy is extreme ($\mathcal{H}^{(1)} = 0$) and the quadratic terms describe collective excitations of the system.

2.1.1.2 The Spectrum of Magnons

The quadratic part of the Hamiltonian can be written as

$$\mathcal{H}^{(2)} = \sum_k \left[\mathcal{A}_k a_k^\dagger a_k + \frac{1}{2}\left(\mathcal{B}_k^* a_k a_{-k} + \mathcal{B}_k a_k^\dagger a_{-k}^\dagger \right) \right] , \tag{2.29}$$

where

$$\begin{aligned} \mathcal{A}_k &= S J(Z_0 - Z_k) + \hbar\gamma \left[H + H_\mathrm{K}\left(1 - \frac{1}{2S}\right)\right] \\ &\quad - S(\hbar\gamma)^2 \left[\sum_r \frac{1}{r^3}\left(1 - 3\frac{z^2}{r^2}\right) - \sum_r \frac{1}{r^3}\left(1 - \frac{3}{2}\frac{r^+ r^-}{r^2}\right) e^{ikr} \right] , \end{aligned} \tag{2.30}$$

$$\mathcal{B}_k = -\frac{3}{2} S(\hbar\gamma)^2 \left[\sum_r \frac{(r^-)^2}{r^5} e^{-ikr} \right] , \tag{2.31}$$

$\mathcal{B}_k^*$ denotes the complex conjugate, and

$$Z_k = \sum_r^{\text{(n.n.)}} \exp(i\boldsymbol{k}\boldsymbol{r}) . \tag{2.32}$$

is the sum over the nearest neighbors.

Using formulas from Appendix B, we can transform dipolar terms as follows:

$$\begin{aligned} -S(\hbar\gamma)^2 &\left[\sum_r \frac{1}{r^3}\left(1 - 3\frac{z^2}{r^2}\right) - \sum_r \frac{1}{r^3}\left(1 - \frac{3}{2}\frac{r^+ r^-}{r^2}\right) e^{ikr} \right] \\ &= -\hbar\gamma 4\pi M_\mathrm{s} \left(N_{zz} - \frac{1}{2}\frac{k_x^2 + k_y^2}{k^2} \right) \\ &= -\hbar\gamma 4\pi M_\mathrm{s} \left(N_{zz} - \frac{1}{2}\sin^2\theta_k \right) , \end{aligned} \tag{2.33}$$

$$\begin{aligned} -\frac{3}{2} S(\hbar\gamma)^2 &\left[\sum_r \frac{(r^+)^2}{r^5} e^{-ikr} \right] \\ &= \frac{\hbar\gamma}{2} M_\mathrm{s} \left[\frac{V_\mathrm{s}}{\mathcal{N}} \sum_r \frac{r^2 - 3x^2}{r^5} e^{ikr} - \frac{V_\mathrm{s}}{\mathcal{N}} \sum_r \frac{r^2 - 3y^2}{r^5} e^{ikr} \right. \\ &\quad \left. -2i \frac{V_\mathrm{s}}{\mathcal{N}} \sum_r \frac{3xy}{r^5} e^{ikr} \right] \\ &= \frac{1}{2}\hbar\gamma 4\pi M_\mathrm{s} \left[\left(\frac{k_x}{k}\right)^2 - \left(\frac{k_y}{k}\right)^2 + 2i\frac{k_x k_y}{k^2} \right] \\ &= 2\pi\hbar\gamma M_\mathrm{s} \left(\frac{k_x + i k_y}{k} \right)^2 \\ &= 2\pi\hbar\gamma M_\mathrm{s} \sin^2\theta_k \exp(i2\phi_k) . \end{aligned} \tag{2.34}$$

From the equations of motion

$$i\hbar\frac{d}{dt}a_k = \left[a_k, \mathcal{H}^{(2)}\right] = \mathcal{A}_k a_k + \mathcal{B}_k a^\dagger_{-k} ,$$
$$i\hbar\frac{d}{dt}a^\dagger_{-k} = \left[a^\dagger_{-k}, \mathcal{H}^{(2)}\right] = -\mathcal{A}_k a^\dagger_{-k} - \mathcal{B}^*_k a_k \tag{2.35}$$

we get the characteristic equation

$$\begin{vmatrix} \mathcal{A}_k - \hbar\omega_k & \mathcal{B}_k \\ \mathcal{B}^*_k & \mathcal{A}_k + \hbar\omega_k \end{vmatrix} = 0 . \tag{2.36}$$

The solution

$$\hbar\omega_k = \operatorname{sign}(\mathcal{A}_k)\sqrt{\mathcal{A}_k^2 - |\mathcal{B}_k|^2} \tag{2.37}$$

is the energy of a magnon in a ferromagnet. In an explicit form the magnon spectrum becomes

$$\omega_k = \gamma\left\{\left[H_\mathrm{i} + H_\mathrm{E}(Z_0 - Z_k)\right]\right.$$
$$\left.\times\left[H_\mathrm{i} + H_\mathrm{E}(Z_0 - Z_k) + 4\pi M_\mathrm{s}\sin^2\theta_k\right]\right\}^{1/2} , \tag{2.38}$$

where $H_\mathrm{i} = H + H_\mathrm{K} - 4\pi M_\mathrm{s} N_{zz}$ is the internal magnetic field and $H_\mathrm{E} = SJ/\hbar\gamma$ is the exchange field. The magnon magnetic moment (2.1), which corresponds to (2.38) is equal to

$$\mu_k = -\hbar\gamma\,\Theta(H_\mathrm{i}, H_\mathrm{E}, M_\mathrm{s}, \boldsymbol{k}) , \tag{2.39}$$

where

$$\Theta(H_\mathrm{i}, H_\mathrm{E}, M_\mathrm{s}, \boldsymbol{k}) = \frac{\gamma[H_\mathrm{i} + H_\mathrm{E}(Z_0 - Z_k) + 2\pi M_\mathrm{s}\sin^2\theta_k]}{\omega_k} .$$

In the long-wavelength approximation $\Theta(.) \approx 1$ and therefore, $\mu_k \approx -\hbar\gamma = -2\mu_\mathrm{B}$, where μ_B is the Bohr magneton.

Applying the linear canonical transformation

$$a_k = u_k b_k + v_k b^\dagger_{-k} ,$$
$$a^\dagger_k = u_k b^\dagger_k + v_k b_{-k} , \tag{2.40}$$

where

$$u_k = \sqrt{\frac{\mathcal{A}_k + \hbar\omega_k}{2\hbar\omega_k}} ,$$
$$v_k = -\frac{\mathcal{B}_k}{|\mathcal{B}_k|}\sqrt{\frac{\mathcal{A}_k - \hbar\omega_k}{2\hbar\omega_k}} , \tag{2.41}$$

to the quadratic form (2.29), we obtain

$$\mathcal{H}^{(2)} = \sum_k \hbar\omega_k b_k^\dagger b_k \,. \tag{2.42}$$

This quadratic Hamiltonian describes an ideal gas of magnons with creation and annihilation operators that satisfy the following relations:

$$\begin{aligned} [b_k, b_q] &= \left[b_k^\dagger, b_q^\dagger\right] = 0\,, \\ \left[b_k, b_q^\dagger\right] &= \Delta(\boldsymbol{k} - \boldsymbol{q})\,. \end{aligned} \tag{2.43}$$

Problem 2.1. Calculate the Hamiltonian of three- and four-magnon interactions.

2.2 Microscopic Modeling

A microscopic approach is another way to describe the magnetic order and excitations in spin systems. In this approach it is assumed that we know all information on crystalline structure, positions of spins and spin–spin interactions. The microscopic approach looks much more accurate and is assumed to be more rigorous for calculations, especially for magnons with large wave vectors close to the boundaries of Brillouin zone. The description of elementary spin excitations (and their interactions) for small wave vectors is supposed to be equivalent both in microscopic and phenomenological approaches. In this section we shall consider an example of the microscopic approach.

2.2.1 Magnons in a Two-Sublattice Antiferromagnet

An overwhelming majority of magnetic systems are antiferromagnets. The local fields in an antiferromagnetic lattice almost compensate for each other. Thus this form of magnetic order significantly differs from the ferromagnetic order, where the local exchange field dominates. Small spin deviations in the antiferromagnet create substantial change of local fields and cause a strong response of the whole spin system. Therefore, the magnetic dynamics in antiferromagnets can exhibit a number of effects that are insignificant in ferromagnets.

Here we shall consider a microscopic model of a two-sublattice antiferromagnet with the "easy-plane" anisotropy and demonstrate specific features of the magnetic dynamics in this system.

2.2.1.1 Hamiltonian

The energy of the antiferromagnetic exchange spin–spin interaction has the form

$$\mathcal{H}_{\text{ex}} = \sum_{g,f} J(|\boldsymbol{r}_{gf}|)(\boldsymbol{S}_g \cdot \boldsymbol{S}_f)\,, \tag{2.44}$$

where $J(|\boldsymbol{r}|) > 0$, is the exchange integral; g and f denote the cell numbers of the "g" and "f" sublattices (the geometrical points $\boldsymbol{r}_g$ and $\boldsymbol{r}_f$, respectively; $\boldsymbol{r}_{gf} \equiv \boldsymbol{r}_g - \boldsymbol{r}_f$).

The Zeeman energy is defined by

$$\mathcal{H}_Z = \hbar\gamma \left(\boldsymbol{H} \cdot \sum_{j=g,f} \boldsymbol{S}_j \right) , \tag{2.45}$$

where $\boldsymbol{H} = (H, 0, 0)$ is the external magnetic field.

$$\mathcal{H}_A = \sum_{g,f} \beta_A(|\boldsymbol{r}_{gf}|) S_g^z S_f^z \tag{2.46}$$

is the energy of the "hard-axis" (or, in other words, "easy-plane") anisotropy ($\beta_A > 0$).

$$\mathcal{H}_{DM} = -\sum_{g,f} D_M(|\boldsymbol{r}_{gf}|)[\boldsymbol{S}_g \times \boldsymbol{S}_f]_z \tag{2.47}$$

is the energy of the Dzyaloshinskii–Moria interaction.

The magnetization of the antiferromagnet is negligibly small compared to the saturation magnetization of the sublattice. Therefore, we can neglect the dipole–dipole energy of the antiferromagnet [23, 24].

If the magnetic atoms have nuclei with nonzero spins, then there exists the so-called hyperfine interaction A_{hf} between the electronic spin $\boldsymbol{S}_j$ and the spin of the nucleus $\boldsymbol{I}_j$ (which is essentially the result of the contact Fermi and dipole–dipole interactions). The nuclear subsystem energy can be written as

$$\mathcal{H}_n = -A_{hf} \sum_{j=g,f} (\boldsymbol{S}_j \cdot \boldsymbol{I}_j) - \hbar\gamma_n \left(\boldsymbol{H} \cdot \sum_{j=g,f} \boldsymbol{I}_j \right) , \tag{2.48}$$

where the second term corresponds to nuclear spin Zeeman energy, and γ_n is the nuclear spin gyromagnetic ratio.

Thus, we can write the Hamiltonian of an antiferromagnet in the following form:

$$\mathcal{H}_{afm} = \mathcal{H}_{ex} + \mathcal{H}_Z + \mathcal{H}_A + \mathcal{H}_{DM} + \mathcal{H}_n . \tag{2.49}$$

The ground state of the spin system is shown in Figure 2.2. The equilibrium systems of coordinates are defined as

$$\begin{aligned} S_g^x &= -S_g^{z1} \sin\theta_S + S_g^{y1} \cos\theta_S , \\ S_g^y &= S_g^{z1} \cos\theta_S + S_g^{y1} \sin\theta_S , \\ S_g^z &= S_g^{x1} , \end{aligned} \tag{2.50}$$

and

$$\begin{aligned} S_f^x &= -S_f^{z2} \sin\theta_S - S_f^{y2} \cos\theta_S , \\ S_f^y &= -S_f^{z2} \cos\theta_S + S_f^{y2} \sin\theta_S , \\ S_f^z &= S_f^{x2} . \end{aligned} \tag{2.51}$$

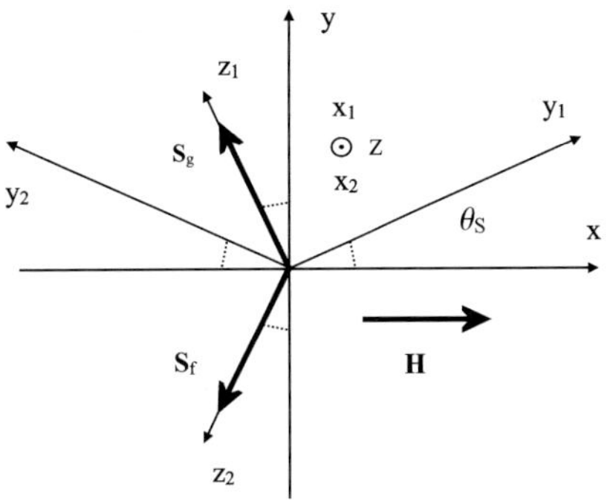

Figure 2.2 Ground state of an "easy-plane" antiferromagnet. $\{x_1, y_1, z_1\}$ and $\{x_2, y_2, z_2\}$ are the axes of quantization.

We can also write the corresponding transformations for the nuclear spins

$$\begin{aligned} I_g^x &= -I_g^{z_3} \sin\theta_I + I_g^{y_3} \cos\theta_I\,, \\ I_g^y &= I_g^{z_3} \cos\theta_I + I_g^{y_3} \sin\theta_I\,, \\ I_g^z &= I_g^{x_3}\,, \end{aligned} \tag{2.52}$$

and

$$\begin{aligned} I_f^x &= -I_f^{z_4} \sin\theta_I - I_f^{y_4} \cos\theta_I\,, \\ I_f^y &= -I_f^{z_4} \cos\theta_I + I_f^{y_4} \sin\theta_I\,, \\ I_f^z &= I_f^{x_4}\,. \end{aligned} \tag{2.53}$$

Now we can represent the spin Hamiltonian (2.49) as the Hamiltonian of nonlinear oscillatory medium using the Holstein–Primakoff transformation:

$$\begin{aligned} S_g^{z_1} &= S - a_g^\dagger a_g\,, \\ S_g^+ &\simeq \sqrt{2S}\,a_g - \frac{a_g^\dagger a_g a_g}{\sqrt{8S}}\,, \\ S_g^- &\simeq \sqrt{2S}\,a_g^\dagger - \frac{a_g^\dagger a_g^\dagger a_g}{\sqrt{8S}}\,, \end{aligned} \tag{2.54}$$

and

$$\begin{aligned} S_f^{z_2} &= S - b_f^\dagger b_f\,, \\ S_f^+ &\simeq \sqrt{2S}\,b_f - \frac{b_f^\dagger b_f b_f}{\sqrt{8S}}\,, \\ S_f^- &\simeq \sqrt{2S}\,b_f^\dagger - \frac{b_f^\dagger b_f^\dagger b_f}{\sqrt{8S}}\,. \end{aligned} \tag{2.55}$$

Here

$$S_g^\pm = S_g^{x_1} \pm i S_g^{y_1}\,, \quad S_f^\pm = S_f^{x_2} \pm i S_f^{y_2}\,,$$

and $a_g^\dagger$, a_g and $b_f^\dagger$, b_f are pairs of the creation and annihilation Bose operators for two spin sublattices.

Now we can represent the initial spin Hamiltonian (2.49) in the form

$$\mathcal{H}_{\text{afm}} = \mathcal{H}^{(0)} + \mathcal{H}^{(1)} + \mathcal{H}^{(2)} + \mathcal{H}^{(3)} + \mathcal{H}^{(4)} + \mathcal{H}_{\text{en}} + \ldots \tag{2.56}$$

Here the upper number indicates the number of normally ordered Bose operators and $\mathcal{H}_{\text{en}}$ denotes the terms, responsible for the nuclear spin deviations.

The energy of the ground state is

$$\begin{aligned}\mathcal{H}^{(0)} &= -\mathcal{N}[(J_0 \cos 2\theta_S + D_{\text{M}0} \sin 2\theta_S) S^2 + 2\hbar\omega_{\text{n}} \langle I \rangle_0 \cos(\theta_S - \theta_I) \\ &\quad + 2H\hbar(\gamma S \sin\theta_S - \gamma_{\text{n}} \langle I \rangle_0 \sin\theta_I)] ,\end{aligned} \tag{2.57}$$

where $\mathcal{N}$ is the number of elementary cells in the sublattice, $\hbar\omega_{\text{n}} = A_{\text{hf}} S$, ω_{n} is the 'unshifted' NMR frequency,

$$J_0 \equiv \sum_{r} J(|\boldsymbol{r}|) , \quad D_{\text{M}0} \equiv \sum_{r} D_{\text{M}}(|\boldsymbol{r}|) ,$$

and $\langle I \rangle_0$ is the mean-field nuclear spin polarization. For $\hbar\omega_{\text{n}} \ll k_{\text{B}} T$ we have

$$\langle I \rangle_0 \simeq \frac{I(I+1)}{3} \frac{\hbar\omega_{\text{n}}}{k_{\text{B}} T} . \tag{2.58}$$

Linear deviations are described by

$$\begin{aligned}\mathcal{H}^{(1)} &= i\sqrt{\frac{S}{2}}[(J_0 \sin 2\theta_S - D_{\text{M}0} \cos 2\theta_S) S - \hbar\gamma H \cos\theta_S + A_{\text{hf}} \langle I \rangle_0 \sin(\theta_S - \theta_I)] \\ &\quad \times \left[\sum_g (a_g - a_g^\dagger) - \sum_f (b_f - b_f^\dagger) \right] \\ &\quad + \frac{i}{2} \left[\hbar\omega_{\text{n}} \sin(\theta_S - \theta_I) - \hbar\gamma_n H \cos\theta_I \right] \\ &\quad \times \left[\sum_g (I_g^+ - I_g^-) - \sum_f (I_f^+ - I_f^-) \right] ,\end{aligned} \tag{2.59}$$

where

$$I_g^\pm = I_g^{x_3} \pm i I_g^{y_3} , \quad I_f^\pm = I_f^{x_4} \pm i I_f^{y_4} .$$

The ground state energy is extreme if $\mathcal{H}^{(1)} = 0$. Thus we obtain

$$\begin{aligned}\theta_S &\simeq \frac{H + H_{\text{DM}}}{2H_{\text{E}}} , \\ \theta_I &\simeq \theta_S - \frac{\gamma_{\text{n}} H}{\omega_{\text{n}}} ,\end{aligned} \tag{2.60}$$

where $H_{\text{DM}} = D_{\text{M}0} S/\hbar\gamma$ is the Dzyaloshinskii–Moria field, and $H_E = J_0 S/\hbar\gamma$ is the exchange field.

Let us now introduce collective variables (and their hermitian conjugates) using a Fourier transformation

$$a_g = \frac{1}{\sqrt{\mathcal{N}}} \sum_k a_k \exp(i\boldsymbol{k}\boldsymbol{r}_g) ,$$
$$b_f = \frac{1}{\sqrt{\mathcal{N}}} \sum_k b_k \exp(i\boldsymbol{k}\boldsymbol{r}_f) . \tag{2.61}$$

2.2.1.2 Spectrum of Magnons

The bosonic quadratic part of the Hamiltonian can be written as

$$\mathcal{H}^{(2)} = \sum_k \left[\mathcal{A}_k \left(a_k^\dagger a_k + b_k^\dagger b_k \right) + \left(\mathcal{B}_k a_k^\dagger b_k + \mathcal{C}_k a_k b_{-k} + \text{h.c.} \right) \right] , \tag{2.62}$$

where h.c. denotes the hermitian conjugate and

$$\mathcal{A}_k = J_0 S \cos 2\theta_S + D_{\text{M}0} S \sin 2\theta_S + \hbar\gamma H \sin\theta_S + A_{\text{hf}} \langle I \rangle_0 \cos(\theta_S - \theta_I) ,$$
$$\mathcal{B}_k = J_0 S \sin^2\theta_S + \frac{1}{2}(\beta_{\text{A}k} S - D_{\text{M}k} S \sin 2\theta_S) ,$$
$$\mathcal{C}_k = J_0 S \cos^2\theta_S + \frac{1}{2}(\beta_{\text{A}k} S + D_{\text{M}k} S \sin 2\theta_S) . \tag{2.63}$$

We have used the following notations:

$$J_k \equiv \sum_r J(|\boldsymbol{r}|) \exp(i\boldsymbol{k}\boldsymbol{r}) ,$$
$$D_{\text{M}k} \equiv \sum_r D_{\text{M}}(|\boldsymbol{r}|) \exp(i\boldsymbol{k}\boldsymbol{r}) ,$$
$$\beta_{\text{A}k} \equiv \sum_r \beta_{\text{A}}(|\boldsymbol{r}|) \exp(i\boldsymbol{k}\boldsymbol{r}) . \tag{2.64}$$

The diagonal form of (2.62)

$$\mathcal{H}^{(2)} = \sum_k \left(\varepsilon_{1k} c_k^\dagger c_k + \varepsilon_{2k} d_k^\dagger d_k \right) \tag{2.65}$$

can be obtained by the linear transformation

$$a_k = \tilde{c}_k + \tilde{d}_k ,$$
$$b_k = \tilde{c}_k - \tilde{d}_k , \tag{2.66}$$

where

$$\tilde{c}_k = U_1(\boldsymbol{k}) c_k + V_1(\boldsymbol{k}) c_{-k}^\dagger ,$$
$$\tilde{d}_k = U_2(\boldsymbol{k}) d_k + V_2(\boldsymbol{k}) d_{-k}^\dagger , \tag{2.67}$$

and

$$
\begin{aligned}
U_1(\mathbf{k}) &= \left(\frac{\mathcal{A}_k + \mathcal{B}_k + \varepsilon_{1k}}{4\varepsilon_{1k}}\right)^{1/2}, \\
V_1(\mathbf{k}) &= -\left(\frac{\mathcal{A}_k + \mathcal{B}_k - \varepsilon_{1k}}{4\varepsilon_{1k}}\right)^{1/2}, \\
U_2(\mathbf{k}) &= \left(\frac{\mathcal{A}_k - \mathcal{B}_k + \varepsilon_{2k}}{4\varepsilon_{2k}}\right)^{1/2}, \\
V_2(\mathbf{k}) &= \left(\frac{\mathcal{A}_k - \mathcal{B}_k - \varepsilon_{2k}}{4\varepsilon_{2k}}\right)^{1/2}.
\end{aligned} \tag{2.68}
$$

The energies of quasi-ferromagnetic and quasi-antiferromagnetic magnons are defined by the formulas

$$
\begin{aligned}
\varepsilon_{1k} &= \sqrt{(\mathcal{A}_k + \mathcal{B}_k)^2 - \mathcal{C}_k^2}, \\
\varepsilon_{2k} &= \sqrt{(\mathcal{A}_k - \mathcal{B}_k)^2 - \mathcal{C}_k^2}.
\end{aligned} \tag{2.69}
$$

The explicit forms for quasi-ferromagnetic and quasi-antiferromagnetic magnon energies are

$$
\begin{aligned}
\varepsilon_{1k} &= \hbar\gamma \sqrt{H(H + H_{\mathrm{DM}}) + H_\Delta^2 + (\alpha k)^2}, \\
\varepsilon_{2k} &= \hbar\gamma \sqrt{2H_E H_A + H_{\mathrm{DM}}(H + H_{\mathrm{DM}}) + H_\Delta^2 + (\alpha k)^2},
\end{aligned} \tag{2.70}
$$

where $H_A = \beta_{A0} S/\hbar\gamma$ is the field of hard-axis anisotropy; $(\hbar\gamma H_\Delta)^2 = 2A_{\mathrm{hf}}\langle I\rangle_0 J_0 S$ is the gap in the magnon spectrum from the hyperfine interaction; $J_k \simeq J_0[1 - (1/2)(\alpha k/J_0 S)^2]$.

Corresponding magnetic moments for quasi-ferromagnetic and quasi-antiferromagnetic magnons are equal to

$$
\begin{aligned}
\mu_{1k} &= -\mu_{\mathrm{B}} \frac{\hbar\gamma(2H + H_{\mathrm{DM}})}{\varepsilon_{1k}}, \\
\mu_{2k} &= -\mu_{\mathrm{B}} \frac{\hbar\gamma H_{\mathrm{DM}}}{\varepsilon_{2k}}.
\end{aligned} \tag{2.71}
$$

As we can see from these relations, the magnetic moments of magnons in an antiferromagnet can significantly differ from $2\mu_{\mathrm{B}}$ [15, 25].

2.2.2 Magnon–Magnon Interactions

Equations (2.67) can be rewritten as

$$
\begin{aligned}
\tilde{c}_k &= \frac{1}{2}\hat{C}_k^- + \frac{1}{4}\hat{C}_k^+, \\
\tilde{d}_k &= \frac{1}{2}\hat{D}_k^+ + \frac{1}{4}\hat{D}_k^-,
\end{aligned} \tag{2.72}
$$

where

$$\hat{C}_k^- \simeq \sqrt{\frac{J}{\varepsilon_{1k}}}(c_k - c_{-k}^\dagger)\,,$$
$$\hat{C}_k^+ \simeq \sqrt{\frac{\varepsilon_{1k}}{J}}(c_k + c_{-k}^\dagger)\,, \tag{2.73}$$

and

$$\hat{D}_k^+ \simeq \sqrt{\frac{J}{\varepsilon_{2k}}}(d_k + d_{-k}^\dagger)\,,$$
$$\hat{D}_k^- \simeq \sqrt{\frac{\varepsilon_{2k}}{J}}(d_k - d_{-k}^\dagger)\,. \tag{2.74}$$

Here $J \equiv J_0 S = \hbar\gamma H_E$ and the factor $\sqrt{J/\varepsilon_{1k}} \gg 1$ in (2.73), the so-called exchange enhancement, plays an extremely important role in magnon dynamics of the antiferromagnet. Mathematically, due to this factor small field perturbations can create a much stronger response in antiferromagnet compared to ferromagnetic systems.

The Hamiltonian of three-magnon interactions can be written as

$$\mathcal{H}^{(3)} = \frac{-i\hbar\gamma H}{\sqrt{2^5 \mathcal{N} S}} \sum_{1,2,3} \left(\hat{C}_1^- \hat{C}_2^- \hat{D}_3^- - 2\hat{C}_1^+ \hat{C}_2^- \hat{D}_3^+ + \hat{D}_1^+ \hat{D}_2^+ \hat{D}_3^- \right) \Delta(\boldsymbol{k}_1 + \boldsymbol{k}_2 + \boldsymbol{k}_3)\,. \tag{2.75}$$

The four-magnon Hamiltonian has the form

$$\begin{aligned}\mathcal{H}^{(4)} = \frac{J_0}{2^5 \mathcal{N}} \sum_{1,2,3,4} \Big[& \hat{C}_1^- \hat{C}_2^- \hat{C}_3^+ \hat{C}_4^+ + \Lambda_1 \hat{C}_1^- \hat{C}_2^- \hat{C}_3^- \hat{C}_4^- \\ & + \hat{C}_1^+ \hat{C}_2^+ \hat{D}_3^+ \hat{D}_4^+ + \hat{C}_1^- \hat{C}_2^- \hat{D}_3^- \hat{D}_4^- \\ & - (\Lambda + \Lambda_0) \hat{C}_1^- \hat{C}_2^- \hat{D}_3^+ \hat{D}_4^+ - 4 \hat{C}_1^+ \hat{C}_2^- \hat{D}_3^+ \hat{D}_4^- \\ & + \hat{D}_1^- \hat{D}_2^- \hat{D}_3^+ \hat{D}_4^+ + (\Lambda_1 + \Lambda_0) \hat{D}_1^+ \hat{D}_2^+ \hat{D}_3^+ \hat{D}_4^+ \Big] \Delta(\boldsymbol{k}_1 + \boldsymbol{k}_2 + \boldsymbol{k}_3 + \boldsymbol{k}_4)\,, \end{aligned} \tag{2.76}$$

where

$$\Lambda_0 = 2\left(\frac{H_A}{H_E}\right) - \left(\frac{H_{DM}}{H_E}\right)^2 + \left(\frac{H}{H_E}\right)^2\,,$$
$$\Lambda_1 = \frac{1}{3}\left(\frac{\alpha k_1}{H_E}\right)^2\,,$$

and

$$\Lambda \equiv \frac{\alpha^2}{H_E^2}\left[k_1^2 + k_3^2 + 2(\boldsymbol{k}_1 \cdot \boldsymbol{k}_2 + \boldsymbol{k}_3 \cdot \boldsymbol{k}_4)\right]\,.$$

We have considered an example demonstrating an analytical representation of the two-sublattice antiferromagnet in the form of two-component magnon gas. Several symmetry properties have been used to simplify calculations. A general diagonalization procedure for two-component Bose systems is given in [26]. It should be noted that the use of symmetry helps to consider analytically a much more complicated system, for example the six-sublattice antiferromagnet [27] and twenty-sublattice ferrite (YIG) [28, 29].

2.3 Nuclear Magnons

A hyperfine interaction (2.48) between electronic and nuclear spins can play an important role in magnetic dynamics. It is convenient to distinguish two components in the hyperfine interaction: a static part, which describes the interaction of the longitudinal magnetizations of the electron and nuclear spins, and a dynamical part, which describes the interaction of their "circular" components. The static hyperfine interaction creates a strong effective magnetic field at the nucleus, which determines the frequency of the free precession of the nuclear spin, the Larmor (or unshifted) nuclear magnetic resonance frequency. The nuclear spin polarized by this field creates a perturbation on the electron shell, which leads to the appearance of a gap in the electronic magnon spectrum. The most interesting physics of collective nuclear and electronic spin dynamics takes place in weakly anisotropic and easy-plane antiferromagnets, such as $CsMnF_3$ and $MnCO_3$. Here we shall briefly discuss this topic.

The term "nuclear spin wave", or "nuclear magnon", denotes the elementary excitation of the coupled electron and nuclear spin oscillations in the nuclear magnetic resonance frequency range [30–34]. The most remarkable property of these excitations is that, at liquid helium temperatures, they are the coupled oscillations of two completely different magnetic subsystems. The electronic spins are ordered by an exchange interaction and the nuclear spins are in a paramagnetic state. As a result of the combined oscillations of these two subsystems, the frequency of the electronic magnons slightly increases, and the nuclear magnetic resonance frequency $\omega_{\mathrm{n}0}$ decreases and becomes noticeably lower than the frequency ω_{n}, which corresponds to the ordinary Larmor precession of nuclear spins. In other words, there arises the dynamic frequency shift of nuclear magnetic resonance, the so-called pulling.

Let us consider a part of the electronic-nuclear spin Hamiltonian which describes the nuclear spin Larmor precession and the linear coupling of the transverse nuclear spin components with quasi-ferromagnetic magnons. It can be written in the form

$$\mathcal{H}'_{\mathrm{en}} = -\hbar\omega_{\mathrm{n}}\hat{m}^z_{\mathrm{n}0} + \frac{\hbar\omega_{\mathrm{n}}}{\sqrt{2^3 S}} \sum_k \hat{C}^-_k \left(\hat{m}^-_{\mathrm{n}-k} - \hat{m}^+_{\mathrm{n}k}\right) \tag{2.77}$$

where

$$\hat{m}^{z}_{\mathrm{n}0} \equiv \sum_{g} I^{z_3}_{g} + \sum_{f} I^{z_4}_{f} \tag{2.78}$$

and

$$\begin{aligned}
\hat{m}^{\pm}_{nk} &\equiv I^{\pm}_{1k} + I^{\pm}_{2k} \,, \\
I^{+}_{1k} &= \frac{1}{\sqrt{\mathcal{N}}} \sum_{g} I^{+}_{g} \exp(-i\boldsymbol{k}\boldsymbol{r}_g) \,, \\
I^{+}_{2k} &= \frac{1}{\sqrt{\mathcal{N}}} \sum_{f} I^{+}_{f} \exp(-i\boldsymbol{k}\boldsymbol{r}_f) \,.
\end{aligned} \tag{2.79}$$

Taking into account the fact that the nuclear Larmor frequency is much less than the quasi-ferromagnetic magnon frequencies, $\omega_{\mathrm{n}} \ll \omega_{1k} = \varepsilon_{1k}/\hbar$, we can exclude the electronic part in the Hamiltonian of the nuclear subsystem. Assuming a condition of quasi equilibrium

$$\frac{\partial}{\partial \psi_k} \left(\mathcal{H}^{(2)} + \mathcal{H}'_{\mathrm{en}}\right) = 0 \,, \tag{2.80}$$

where

$$\psi_k \equiv \hat{C}^{-}_{k}$$

is a generalized coordinate, one obtains

$$\hat{C}^{-}_{k} \simeq \frac{J_0 S}{\varepsilon^2_{1k}} \frac{\hbar\omega_{\mathrm{n}}}{\sqrt{2S}} \left(\hat{m}^{-}_{\mathrm{n}-k} - \hat{m}^{+}_{\mathrm{n}k}\right) . \tag{2.81}$$

This relation simply reflects how slow nuclear spin motion is followed by motion of electronic magnons.

Now we can write the effective nuclear spin Hamiltonian in the form

$$\mathcal{H}_{\mathrm{nn}} = -\hbar\omega_{\mathrm{n}}\hat{m}^{z}_{\mathrm{n}0} + \sum_{k} \frac{J_0}{4} \frac{(AS)^2}{\varepsilon^2_{1k}} \left(\hat{m}^{-}_{n-k} - \hat{m}^{+}_{nk}\right)\left(\hat{m}^{-}_{nk} - \hat{m}^{+}_{n-k}\right) . \tag{2.82}$$

Here, the second term is known as an indirect Suhl–Nakamura interaction [35, 36]. It represents a kind of XY spin model with possible phase transitions [37].

From the linearized equations of motion

$$\begin{aligned}
i\hbar \frac{d}{dt} \hat{m}^{-}_{k} &= \left[\hat{m}^{-}_{k}, \mathcal{H}_{\mathrm{nn}}\right] , \\
i\hbar \frac{d}{dt} \hat{m}^{+}_{k} &= \left[\hat{m}^{+}_{k}, \mathcal{H}_{\mathrm{nn}}\right] ,
\end{aligned} \tag{2.83}$$

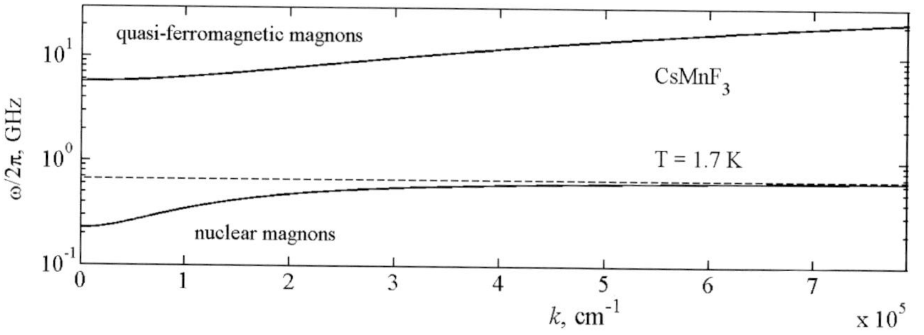

Figure 2.3 The spectra of electronic and nuclear magnons.

one can obtain the spectrum of nuclear magnons

$$\omega_{nk} = \omega_n \sqrt{1 - \left(\frac{\gamma H_\Delta}{\omega_{1k}}\right)^2} \,. \tag{2.84}$$

Figure 2.3 shows a typical spectra of electronic and nuclear magnons in $CsMnF_3$.

The magnetic moment of a nuclear magnon can be represented as

$$\mu_{nk} = -\frac{\omega_n}{\omega_{nk}} \frac{\omega_n}{\omega_{1k}} \left(\frac{\gamma H_\Delta}{\omega_{1k}}\right)^2 \mu_{1k} \,. \tag{2.85}$$

Problem 2.2. Estimate the magnetic moments of electronic (quasi-ferromagnetic) and nuclear magnons in $MnCO_3$ for the following conditions: $\omega_n/2\pi = 640\,\text{MHz}$, $T = 1.7\,\text{K}$, $H = 39.9\,\text{kA/m}$, $H_{DM} = 351\,\text{kA/m}$, $H_\Delta^2 = 36\,731/T\,[\text{K}]\,(\text{A/m})^2$. Magnetic moments of nuclear magnons were studied experimentally in [38].

2.4
Magnetoelastic Waves, Quasi Phonons

The elementary excitations of the ordered magnetic system are magnons, and those of the elastic solid are phonons. Magnetoelastic interactions give rise to a coupling of magnetic and elastic oscillations. The resulting normal modes (quasi magnons and quasi phonons) contain both the magnetic and elastic components. This means that such quasi particles can be excited, in principle, either by an alternating magnetic field or by elastic vibrations. Here, we consider quasi phonons in the easy-plane antiferromagnet (such as $FeBO_3$, hematite, and so on), where the magnetoelastic phenomena exhibit so-called exchange enhancement (see, for example, [39]).

The Hamiltonian of the two-sublattice antiferromagnet can be represented as

$$\mathcal{H} = \mathcal{H}_{\text{afm}} + \mathcal{H}_{\text{e}} + \mathcal{H}_{\text{me}} \,, \tag{2.86}$$

where

$$\mathcal{H}_{\mathrm{e}} + \mathcal{H}_{\mathrm{me}} = \int (\mathcal{U}_{\mathrm{e}} + \mathcal{U}_{\mathrm{me}}) d\boldsymbol{r} \tag{2.87}$$

are the elastic and magnetoelastic energies of the sample.

The explicit forms of elastic and magnetoelastic energies per unit volume for an antiferromagnet with rhombohedral symmetry can be written as (see, for example, [40])

$$\begin{aligned}\mathcal{U}_{\mathrm{e}} &= \frac{1}{2}\rho(\dot{\boldsymbol{U}})^2 + \frac{1}{2}\mathcal{C}_{11}(u_{xx}^2 + u_{yy}^2) + \frac{1}{2}\mathcal{C}_{33}u_{zz}^2 + \mathcal{C}_{12}u_{xx}u_{yy}\\ &+ \mathcal{C}_{13}(u_{xx} + u_{yy})u_{zz} + (\mathcal{C}_{11} - \mathcal{C}_{12})u_{xy}^2 + 2\mathcal{C}_{44}(u_{yz}^2 + u_{xz}^2)\\ &+ 2\mathcal{C}_{14}[(u_{xx} - u_{yy})u_{yz} + 2u_{xy}u_{xz}] ,\end{aligned} \tag{2.88}$$

and

$$\begin{aligned}\mathcal{U}_{\mathrm{me}} &= \mathcal{B}_{11}\left(l_x^2 u_{xx} + l_y^2 u_{yy}\right) + \mathcal{B}_{12}\left(l_x^2 u_{yy} + l_y^2 u_{xx}\right) + \mathcal{B}_{33}u_{zz}l_z^2\\ &+ 2\mathcal{B}_{44}(u_{yz}l_y + u_{xz}l_x)l_z + 2(\mathcal{B}_{11} - \mathcal{B}_{12})u_{xy}l_x l_y\\ &+ 2\mathcal{B}_{14}\left[\left(l_x^2 - l_y^2\right)u_{yz} + 2l_x l_y u_{xz}\right]\\ &+ \mathcal{B}_{41}[l_x l_z(u_{xx} - u_{yy}) + 2l_x l_z u_{xy}] .\end{aligned} \tag{2.89}$$

Here,

$$u_{ij} = \frac{1}{2}\left(\frac{\partial U_i}{\partial x_j} + \frac{\partial U_j}{\partial x_i}\right) \tag{2.90}$$

is the strain tensor, $\boldsymbol{U}$ the displacement vector, ρ is a density, $\mathcal{C}_{ij}$ the elastic modulus tensor [41], and $\mathcal{B}_{ij}$ the magnetoelastic tensor.

The antiferromagnetic vector $\boldsymbol{l} = (l_x, l_y, l_z)$ can be expressed as

$$\boldsymbol{l} = \frac{\boldsymbol{M}_1 - \boldsymbol{M}_2}{2M_{\mathrm{s}}} = \frac{\boldsymbol{S}_g - \boldsymbol{S}_f}{2S} , \tag{2.91}$$

where $|\boldsymbol{M}_\nu| = M_{\mathrm{s}}$ is the magnetization of the sublattice.

Taking into account considered above transformations for electronic spins into magnons, we can obtain the following relations for the quasi-ferromagnetic oscillations:

$$\begin{aligned} l_x &\simeq \frac{-i}{\sqrt{2\mathcal{N}S}}\sum_{\boldsymbol{k}}\exp(i\boldsymbol{k}\boldsymbol{r})\hat{C}_k^- ,\\ l_y &\simeq 1 + \frac{1}{4\mathcal{N}S}\sum_{\boldsymbol{k},\boldsymbol{q}}\exp\left[i(\mathbf{k}+\mathbf{q})\mathbf{r}\right]\hat{C}_k^-\hat{C}_q^- ,\\ l_z &\simeq 0 .\end{aligned} \tag{2.92}$$

Indirect Elastic–Elastic Interactions

In the region of acoustic frequencies which is far below the frequencies of quasi-ferromagnetic magnons, the magnetoelastic interactions can be simplified by introducing effective elastic–elastic interactions [42]. We shall assume conditions of quasi equilibrium of the form

$$\frac{\partial\left(\mathcal{H}^{(2)}+\mathcal{H}_{\mathrm{me}}\right)}{\partial\psi_{k}}=0\,, \tag{2.93}$$

where

$$\psi_{k}\equiv\hat{C}_{k}^{-}$$

is the generalized coordinate. We thereby exclude the magnetic degrees of freedom and obtain the following relation:

$$\begin{aligned}\hat{C}_{k}^{-}\simeq{}&-iV_{\mathrm{s}}\sqrt{\frac{2S}{\mathcal{N}^{3}}}\sum_{r}\frac{2J_{0}}{\varepsilon_{1k}^{2}}e^{-ikr}[(\mathcal{B}_{11}-\mathcal{B}_{12})u_{xy}(r)+2\mathcal{B}_{14}u_{xz}(r)]\\&\times\left\{1-\sum_{q}\sum_{r_{1}}\frac{2J_{0}}{\varepsilon_{1k}^{2}}e^{-iq(r_{1}-r)}[(\mathcal{B}_{11}-\mathcal{B}_{12})u_{xy}(r_{1})+2\mathcal{B}_{14}u_{xz}(r_{1})]\right\}\,.\end{aligned} \tag{2.94}$$

Eliminating the magnetic degrees of freedom we obtain the effective elastic–elastic interactions. For example, the second-order term in u_{ij} has the form

$$\tilde{\mathcal{U}}_{\mathrm{e}}=\mathcal{U}_{\mathrm{e}}+\Delta\mathcal{U}_{\mathrm{e}}\,,$$

where

$$\begin{aligned}\Delta\mathcal{U}_{\mathrm{e}}(r)={}&-\frac{1}{\mathcal{N}}\sum_{k}\sum_{r_{1}}\frac{2J_{0}}{\varepsilon_{fk}^{2}}e^{-ik(r_{1}-r)}[(\mathcal{B}_{11}-\mathcal{B}_{12})u_{xy}(r)+2\mathcal{B}_{14}u_{xz}(r)]\\&\times[(\mathcal{B}_{11}-\mathcal{B}_{12})u_{xy}(r_{1})+2\mathcal{B}_{14}u_{xz}(r_{1})]\end{aligned} \tag{2.95}$$

describes the indirect elasticity mediated by the magnetic subsystem. Note that this elastic energy term becomes nonlocal.

The spectrum $\Omega_{\mathrm{me}k}$ and polarization $e(k,s)$ of magnetoelastic modes of the system are determined by the Green–Christoffel equation

$$[\rho\Omega_{\mathrm{me}k}I-\hat{\Gamma}(k)]e(k,s)=0\,, \tag{2.96}$$

which is obtained after going over to the plane wave representation in the dynamic equation for the displacement U.

One of the most interesting objects used to study the magnetic excitation of quasi phonons is the high Néel temperature antiferromagnet iron borate ($FeBO_3$,

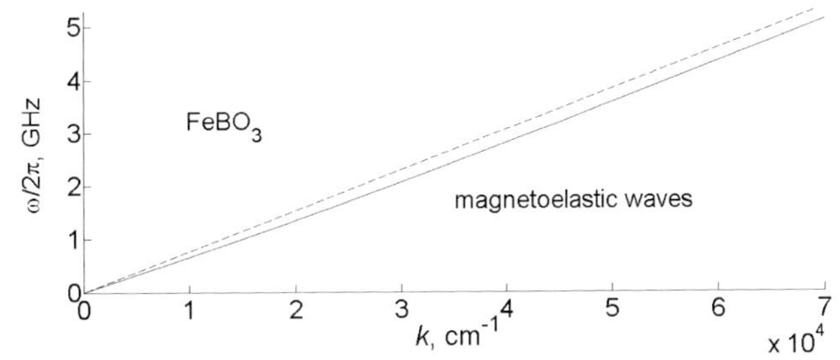

Figure 2.4 The spectrum of quasi phonons.

$T_N = 348$ K). It has rhombohedral crystalline symmetry with the C_3-axis perpendicular to the "easy plane" and exhibits strong magnetoelastic dynamics. The magnetic moments of quasi phonons in this crystal have been studied in [43]. Later, in Chapter 4 we will focus on the transverse quasi phonons (with the wave vectors in the C_3 direction and polarization parallel to the static magnetic field), which are the most strongly coupled to the quasi-ferromagnetic magnons. The spectrum of these quasi phonons (see Figure 2.4) can be written as

$$\omega_k = c_e k \cdot \left[1 - \frac{(H_{\Delta me}\xi)^2}{H(H + H_{DM}) + H_{\Delta me}^2 + (\alpha k)^2} \right]^{1/2} . \tag{2.97}$$

Here, $c_e \simeq 4.8 \times 10^3$ m/s is the sound velocity, $H_{\Delta me}\xi \simeq 2$ kOe (159 kA/m) the parameter describing an efficiency of the linear interaction between spin and elastic subsystems, $H_{DM} \simeq 100$ kOe (7.96×10^6 A/m) is the Dzyaloshinskii field, $H_{\Delta me} \simeq 2.2$ kOe (175 kA/m) is the field which corresponds to the magnetoelastic gap in the spin wave spectrum, and $\alpha \simeq 0.08$ Oe cm (0.0637 A) the phenomenological exchange constant, which is proportional to the exchange field $H_E = 2.6 \times 10^6$ Oe (2.07×10^8 A/m) (see also, [44]).

2.5 Discussion

a) What are specific features of the phenomenological approach for the magnetic structures?
b) Why is the dipole–dipole interaction less important in an antiferromagnet?
c) What is the most remarkable property of nuclear magnons?
d) Why can quanta of magnetoelastic waves (quasi phonons) be excited by an alternating magnetic field?

3
Relaxation of Magnons

Magnons have a finite lifetime. The lifetime depends on the relaxation processes due to interaction of magnons with each other and with other quasi particles. The magnon–magnon processes characterize the energy exchange inside the magnon system. The magnon–phonon and other similar processes describe the energy transfer to a thermal bath. The relaxation can also depend on the process of interaction of magnons with damped subsystems that absorb the magnon energy and transfers it to the thermal bath. The aim of this chapter is to consider several examples of magnon relaxation processes.

3.1
Master Equation

We shall start with a general approach to the problem of relaxation in quantum systems which was developed in the theory of lasers [5, 45, 46]. The approach is based on the so-called master equation. The master equation is a convenient approximate form of the density matrix equation. Here we shall describe this approach.

The evolution of quantum system is described by the density matrix equation (Liouville–von Neumann equation):

$$i\hbar\frac{\partial\rho}{\partial t} = [\mathcal{H}, \rho] \,. \tag{3.1}$$

We can consider the Hamiltonian of the system $\mathcal{H}$ in the following form

$$\mathcal{H} = \mathcal{H}_\mathrm{d} + \mathcal{H}_\mathrm{b} + \mathcal{V} \,, \tag{3.2}$$

where $\mathcal{H}_\mathrm{d}$ is the Hamiltonian of the dynamic system of interest (for example one magnon), $\mathcal{H}_\mathrm{b}$ is the Hamiltonian of the thermal bath, and $\mathcal{V}$ describes a weak interaction between the dynamic system and thermal bath. In general, this interaction can be represented as

$$\mathcal{V} = \sum_j Q_j F_j \,, \tag{3.3}$$

Nonequilibrium Magnons, First Edition. Vladimir L. Safonov.

where Q_j and F_j correspond to the operators of the dynamic system and thermal bath, respectively.

The density matrix of a thermal bath with the temperature T is defined by

$$\rho_b = \frac{\exp(-\mathcal{H}_b/(k_B T))}{\mathrm{Tr}[\exp(-\mathcal{H}_b/(k_B T))]} , \tag{3.4}$$

and the thermal bath operators are assumed to have zero averages:

$$\langle F_j \rangle_b \equiv \mathrm{Tr}(F_j \rho_b) = 0 . \tag{3.5}$$

We can represent the density matrix in the equation (3.1) in the form

$$\rho = \rho_d \cdot \rho_b + \Delta\rho , \tag{3.6}$$

where $\rho_d = \mathrm{Tr}_b(\rho)$ is the density matrix of dynamic system, and $\mathrm{Tr}_b(\cdot)$ denotes the trace over thermal bath states. From the density matrix equation (3.1) we can get the system of equations for ρ_d and $\Delta\rho$.

Solving the equation for $\Delta\rho$ by means of iterations, one obtains the equation for ρ_d, which is valid up to the second order with respect to $\mathcal{V}$:

$$\frac{d\rho_d}{dt} = -\frac{i}{\hbar}[\mathcal{H}_d, \rho_d] - \frac{1}{\hbar^2} \int_{-\infty}^{t} d\tau \mathrm{Tr}\left\{[V, [V(\tau, t), \rho_d(t, \tau)\rho_b]]\right\} , \tag{3.7}$$

where

$$\mathcal{V}(\tau, t) = \sum_j Q_j(\tau - t) F_j(\tau - t) , \tag{3.8}$$

$$Q_j(u) = \exp\left(\frac{i\mathcal{H}_d u}{\hbar}\right) Q_j \exp\left(-\frac{i\mathcal{H}_d u}{\hbar}\right) , \tag{3.9}$$

$$F_j(u) = \exp\left(\frac{i\mathcal{H}_b u}{\hbar}\right) F_j \exp\left(-\frac{i\mathcal{H}_b u}{\hbar}\right) , \tag{3.10}$$

correspond to interaction representation of Q_j and F_j, and

$$\rho_d(t, \tau) = \exp[-i\mathcal{H}_d(t - \tau)]\rho_d(\tau) \exp[i\mathcal{H}_d(t - \tau)] . \tag{3.11}$$

One can assume that the characteristic times for the $\langle F_j(0) F_j(\tau)\rangle_b$ correlators are much smaller than the characteristic time of ρ_d evolution. Then in (3.7) we can approximately write

$$\rho_d(t, \tau) \approx \rho_d(t, 0) . \tag{3.12}$$

This relation substantially simplifies (3.7) from which one can obtain the equation for the mean value of any operator of dynamical system $\langle \mathcal{O} \rangle \equiv \mathrm{Tr}(\mathcal{O} \cdot \rho_d)$:

$$\begin{aligned} \frac{d\langle \mathcal{O} \rangle}{dt} &= -\frac{i}{\hbar}\langle[\mathcal{O}, \mathcal{H}_d]\rangle \\ &\quad - \frac{1}{\hbar^2} \int_0^{\infty} du \sum_{l,j} \{\langle F_l(u) F_j(0)\rangle_b \langle[\mathcal{O}, Q_l] Q_j(-u)\rangle \\ &\quad - \langle F_j(0) F_l(u)\rangle_b \langle Q_j(-u)[\mathcal{O}, Q_l]\rangle\} . \end{aligned} \tag{3.13}$$

3.2 Relaxation of Bose Quasi Particles

Now we consider the Hamiltonian, which corresponds to Bose quasi particles (e.g., magnons) with the wave vector $\boldsymbol{k}$ and energy $\hbar\omega_k$:

$$\mathcal{H}_\mathrm{d} = \hbar\omega_k b_k^\dagger b_k \, . \tag{3.14}$$

The interaction with a thermal bath has the form

$$\mathcal{V} = b_k F_k^\dagger + b_k^\dagger F_k \, . \tag{3.15}$$

Let us consider the case, when $\mathcal{O} = b_k$. Taking into account that

$$\begin{aligned} b_k(u) &= \exp\left(\frac{i\mathcal{H}_\mathrm{d} u}{\hbar}\right) b_k \exp\left(-\frac{i\mathcal{H}_\mathrm{d} u}{\hbar}\right) = b_k \exp(-i\omega_k u) \, , \\ b_k^\dagger(u) &= \exp\left(\frac{i\mathcal{H}_\mathrm{d} u}{\hbar}\right) b_k^\dagger \exp\left(-\frac{i\mathcal{H}_\mathrm{d} u}{\hbar}\right) = b_k^\dagger \exp(i\omega_k u) \, , \end{aligned} \tag{3.16}$$

from (3.13) we obtain the equation for the averaged b_k:

$$\frac{d\,\langle b_k\rangle}{dt} + \eta_k \langle b_k\rangle = -i(\omega_k + \Delta\omega_k)\langle b_k\rangle \, . \tag{3.17}$$

Here the relaxation rate η_k and the frequency shift $\Delta\omega_k$ are defined by the real and imaginary part of the thermal bath correlator $\langle [F_k(u), F_k^\dagger(0)]\rangle_\mathrm{b}$:

$$\eta_k \equiv \frac{1}{\hbar^2} \operatorname{Re}\left\{\int_0^\infty du \langle [F_k(u), F_k^\dagger(0)]\rangle_\mathrm{b} \exp(i\omega_k u)\right\} , \tag{3.18}$$

$$\Delta\omega_k \equiv \frac{1}{\hbar^2} \operatorname{Im}\left\{\int_0^\infty du \langle [F_k(u), F_k^\dagger(0)]\rangle_\mathrm{b} \exp(i\omega_k u)\right\} . \tag{3.19}$$

The "nondiagonal" bath correlators $\langle [F_k(u), F_k(0)]\rangle_\mathrm{b}$ and $\langle [F_k^\dagger(u), F_k^\dagger(0)]\rangle_\mathrm{b}$ are assumed to be equal to zero to exclude nondiagonal term $\langle b_k^\dagger\rangle$ in (3.17).

Analogously we can obtain the equation for $\mathcal{O} = b_k^\dagger b_k$:

$$\frac{d\left\langle b_k^\dagger b_k\right\rangle}{dt} + 2\eta_k\left(\left\langle b_k^\dagger b_k\right\rangle - \bar{n}_k\right) = 0 \, . \tag{3.20}$$

Here, $\bar{n}_k$ denotes the equilibrium population of magnons.

3.2.1 Relaxation Process of Harmonic Oscillators

Consider now the thermal bath Hamiltonian as a set of harmonic oscillators

$$\mathcal{H}_\mathrm{b} = \hbar\sum_q \Omega_q d_q^\dagger d_q \tag{3.21}$$

with the spectrum Ω_q. These oscillators could be magnons, phonons and other elementary Bose excitations.

In this case the interaction representation for $d_k^\dagger$ and d_k Bose operators is

$$\exp\left(\frac{i\mathcal{H}_b u}{\hbar}\right) d_k \exp\left(-\frac{i\mathcal{H}_b u}{\hbar}\right) = d_k \exp(-i\Omega_k u) ,$$
$$\exp\left(\frac{i\mathcal{H}_b u}{\hbar}\right) d_k^\dagger \exp\left(-\frac{i\mathcal{H}_b u}{\hbar}\right) = d_k^\dagger \exp(i\Omega_k u) . \tag{3.22}$$

As an example we take the bath operator in the form

$$F_k = \sum_{k_1,k_2} \Psi(k; k_1, k_2) d_{k_1} d_{k_2} \Delta(k - k_1 - k_2) . \tag{3.23}$$

It describes the decay of the boson with energy $\hbar\omega_k$ into two elementary excitations with energies $\hbar\Omega_{k_1}$ and $\hbar\Omega_{k_2}$, correspondingly.

The interaction representation (3.10) for F_k can be obtained with the use of (3.22):

$$F_k(u) = \sum_{k_1,k_2} \Psi(k; k_1, k_2) d_{k_1} d_{k_2} e^{-i(\Omega_{k_1}+\Omega_{k_2})u} \Delta(k - k_1 - k_2) . \tag{3.24}$$

Utilizing the commutation

$$\left\langle [d_{k_3} d_{k_4}, d_{k_1}^\dagger d_{k_2}^\dagger] \right\rangle_b = \left\langle d_{k_3} d_{k_4} d_{k_1}^\dagger d_{k_2}^\dagger \right\rangle_b - \langle d_{k_1}^\dagger d_{k_2}^\dagger d_{k_3} d_{k_4} \rangle_b \tag{3.25}$$

and the following rule of decoupling

$$\langle d_{k_1}^\dagger d_{k_2}^\dagger d_{k_3} d_{k_4} \rangle_b \simeq \langle d_{k_1}^\dagger d_{k_1} \rangle_b \langle d_{k_2}^\dagger d_{k_2} \rangle_b \times [\Delta(k_1 - k_3)\Delta(k_2 - k_4) + \Delta(k_1 - k_4)\Delta(k_2 - k_3)] , \tag{3.26}$$

from (3.18) and (3.19) we obtain

$$\eta + i\Delta\omega = -\frac{2}{\hbar^2} \int du \sum_{k_1,k_2} |\Psi(k; k_1, k_2)|^2 \times [(n_{k_1} + 1)(n_{k_2} + 1) - n_{k_2} n_{k_2}] \times \exp[iu(\omega_k - \Omega_{k_1} - \Omega_{k_2})]\Delta(k - k_1 - k_2) . \tag{3.27}$$

Here,

$$n_q \equiv \langle d_q^\dagger d_q \rangle_b = \frac{1}{\exp(\hbar\Omega_q / k_B T) - 1} \tag{3.28}$$

is the boson occupation number.

Taking into account that

$$\int_0^\infty du \exp(i\Omega u) = \pi\delta(\Omega) + iP(1/\Omega) , \tag{3.29}$$

we can rewrite the expression (3.27) in the form:

$$\begin{aligned}\eta_k = &\frac{2\pi}{\hbar}\sum_{k_1,k_2} |\Psi(k;k_1,k_2)|^2 \\ &\times [(n_{k_1}+1)(n_{k_2}+1) - n_{k_2}n_{k_2}] \\ &\times \delta(\hbar\omega_k - \hbar\Omega_{k_1} - \hbar\Omega_{k_2})\Delta(k - k_1 - k_2) ,\end{aligned} \tag{3.30}$$

which corresponds to Fermi's golden rule, that is, the second order of perturbation theory (see, for example, [47]).

Problem 3.1.

a) Derive an explicit formula for the relaxation rate for the case of bath operator, which corresponds to the process of confluence:

$$F = \sum_{k_1,k_2} \Psi(k;k_1,k_2) d^\dagger_{k_1} d_{k_2} \Delta(k + k_1 - k_2) . \tag{3.31}$$

b) Derive an explicit formula for the relaxation rate for the case of bath operator, which corresponds to the four-quasi particle interaction:

$$F = \sum_{k_1,k_2,k_3} \Phi(k,k_1;k_2,k_3) d^\dagger_{k_1} d_{k_2} d_{k_3} \Delta(k + k_1 - k_2 - k_3) . \tag{3.32}$$

Analysis of magnon relaxation processes is an important part of magnetic physics. Comparison of the theory with experimental data makes it possible to understand the role of particular interactions and processes in magnetic dynamics (see, e.g., [48, 49]).

3.2.2
Magnon–Electron Scattering

Now we shall consider the magnon interaction with conduction electrons in a ferromagnetic metal. For simplicity our analysis will be restricted to the case $\boldsymbol{k} = 0$, which corresponds to magnons of uniform magnetization precession with the frequency ω_0. In other words, we are going to analyze the effect of conduction electrons on the linewidth of ferromagnetic resonance.

From a microscopic point of view the most probable interaction process is the confluence of a magnon (with wave vector $\boldsymbol{k} = 0$ and energy $\hbar\omega_0$) and a conduction electron with wave vector $\boldsymbol{q} \neq 0$ and energy ϵ_q into a conduction electron with wave vector $\boldsymbol{q}'$ and energy $\epsilon_{q'} = \epsilon_q + \hbar\omega_0$. It is obvious that this process is forbidden for an ideal crystal, where the momentum conservation is valid $\boldsymbol{q}' = \boldsymbol{q} + \boldsymbol{0}$.

However, the confluence process can occur in the presence of defects, impurities or fluctuations which permit violation of momentum conservation in the crystal (see, Figure 3.1).

We consider small-amplitude magnetization motions of a ferromagnet in the vicinity of equilibrium state $\boldsymbol{M} = M_s\hat{\boldsymbol{z}}$. Here $\hat{\boldsymbol{z}}$ is the unit vector in the equilibrium direction and M_s is the saturation magnetization. The magnetization rotation around the effective field in this case, in general, is elliptical and the magnetic energy can be represented in a quadratic form:

$$\frac{\mathcal{E}}{V} = \frac{H_x}{2M_s}M_x^2 + \frac{H_y}{2M_s}M_y^2 . \tag{3.33}$$

Here, V is the volume of the sample, H_x and H_y are positive "stiffness" fields, which include both microscopic and shape anisotropies and the external magnetic field.

It is convenient to describe small oscillations of the magnetization in terms of creation $a^\dagger$ and annihilation a Bose operators introduced by a linearized Holstein–Primakoff transformation. For the transverse magnetization components one can write

$$M_x \simeq -\sqrt{\frac{\hbar\gamma M_s}{2V}}(a^\dagger + a) ,$$
$$M_y \simeq -\frac{1}{i}\sqrt{\frac{\hbar\gamma M_s}{2V}}(a^\dagger - a) . \tag{3.34}$$

The interaction of the uniform magnetic precession (in terms of a and $a^\dagger$) with electrons is assumed to be of the form [51]:

$$\mathcal{V} = \frac{D^\dagger a + D a^\dagger}{\sqrt{N}} , \tag{3.35}$$

where

$$D = \frac{1}{N}\sum_{q,q'}\sum_{j} f_{q,q'}(\boldsymbol{r}_j)d_{q'}^\dagger d_q . \tag{3.36}$$

The amplitude

$$f_{q,q'}(\boldsymbol{r}_j) = |f_{qq'}|\exp[i\phi(\boldsymbol{r}_j)] \tag{3.37}$$

describes the scattering process in the vicinity of crystal defect at the point $\boldsymbol{r}_j$; $\phi(\boldsymbol{r}_j)$ is the phase. The terms $d_q^\dagger$ and d_q are the fermion creation and annihilation

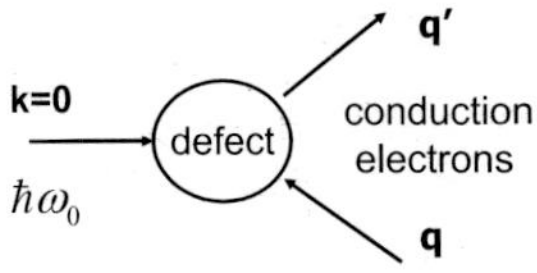

Figure 3.1 Relaxation process of confluence of magnon and conduction electron into conduction electron in the presence of a defect.

operators, respectively,

$$\begin{aligned} d_k^\dagger d_q + d_q d_k^\dagger &= \Delta(\boldsymbol{k} - \boldsymbol{q}) , \\ d_k d_q + d_q d_k &= 0 , \\ d_k^\dagger d_q^\dagger + d_q^\dagger d_k^\dagger &= 0 , \end{aligned} \tag{3.38}$$

describing a conduction electron with wave number $\boldsymbol{q}$ and energy ϵ_q.

The energy (3.33) can be rewritten in the quadratic form:

$$\frac{\mathcal{E}}{\hbar} = \mathcal{A} a^\dagger a + \frac{\mathcal{B}}{2}(aa + a^\dagger a^\dagger) , \tag{3.39}$$

where

$$\mathcal{A} = \frac{\gamma(H_x + H_y)}{2} \text{ and } \mathcal{B} = \frac{\gamma(H_x - H_y)}{2} . \tag{3.40}$$

The canonical transformation of (3.39) to a normal mode

$$\mathcal{E} = \hbar\omega_0 b^\dagger b \tag{3.41}$$

with frequency

$$\omega_0 = \sqrt{\mathcal{A}^2 - \mathcal{B}^2} = \gamma \sqrt{H_x H_y} \tag{3.42}$$

and Bose operators $b^\dagger$ and b is performed by

$$\begin{aligned} a &= ub + vb^\dagger , \\ a^\dagger &= ub^\dagger + vb . \end{aligned} \tag{3.43}$$

The coefficients

$$u = \sqrt{\frac{\mathcal{A} + \omega_0}{2\omega_0}} , \quad v = -\frac{\mathcal{B}}{|\mathcal{B}|}\sqrt{\frac{\mathcal{A} - \omega_0}{2\omega_0}} . \tag{3.44}$$

describe ellipticity of the magnetization rotation, which depends on a concrete form of the magnetic Hamiltonian.

In terms of the normal mode the Hamiltonian (3.35) becomes

$$\mathcal{V} = \frac{1}{N^{3/2}} \sum_{k,k'} \sum_{j} \left[\Psi_{kk'}(\boldsymbol{r}_j) d_{k'}^\dagger d_k b + \Psi_{kk'}^*(\boldsymbol{r}_j) d_k^\dagger d_{k'} b^\dagger \right] , \tag{3.45}$$

where

$$\Psi_{q,q'}(\boldsymbol{r}_j) = u_k f_{q,q'}(\boldsymbol{r}_j) + v_k f_{q,q'}^*(\boldsymbol{r}_j) \tag{3.46}$$

is the amplitude of magnon–electron scattering.

The formula for the relaxation rate can be obtained from (3.13) with a thermal bath of conduction electrons, and is defined by Fermi's golden rule:

$$\begin{aligned}\eta_{\text{m-el}} &= \frac{\pi}{N^3\hbar}\sum_{q,q'}\sum_{j}|\Psi_{q',q}|^2(\overline{n}_q - \overline{n}_{q'})\delta(\epsilon_{q'} - \epsilon_q - \hbar\omega_k) \\ &= \frac{\pi}{N^3\hbar}\sum_{q,q'}\sum_{j}\{u^2|f_{q',q}|^2 + v^2|f_{q,q'}|^2 \\ &\quad + uv|f_{q',q}f_{q,q'}|\cos[2\phi(\boldsymbol{r}_j)]\} \\ &\quad \times(\overline{n}_q - \overline{n}_{q'})\delta(\epsilon_{q'} - \epsilon_q - \hbar\omega_k)\,. \end{aligned} \tag{3.47}$$

Here,

$$\overline{n}_q \equiv \langle d_q^\dagger d_q\rangle_{\text{b}} = \frac{1}{\exp[(\epsilon_q - \epsilon_{\text{F}})/k_{\text{B}}T] + 1} \tag{3.48}$$

is the Fermi occupation number, where ϵ_{F} is the Fermi energy.

If we assume that the scattering phases $\phi(\boldsymbol{r}_j)$ are random and $|f_{k'k}| = |f_{kk'}|$, then the uv term in (3.47) vanishes and the relaxation rate becomes

$$\begin{aligned}\eta_{\text{m-el}} &= \frac{c_{\text{def}}\pi(u^2 + v^2)}{N^2\hbar}\sum_{q,q'}|f_{q,q'}|^2 \\ &\quad \times(\overline{n}_q - \overline{n}_{q'})\delta(\epsilon_{q'} - \epsilon_q - \hbar\omega_0)\,. \end{aligned} \tag{3.49}$$

Here, $c_{\text{def}} = N_{\text{def}}/N$ is the concentration of defects, and N_{def} is the total number of defects in the crystal.

After the following simplifications, $\overline{n}_q - \overline{n}_{q'} \simeq -\hbar\omega_0\partial\overline{n}_q/\partial\epsilon_q$ and, assuming that the magnon energy is much less than the conduction electron energy $\hbar\omega_0 \ll \epsilon_q$, we can obtain [51]

$$\eta_{\text{m-el}} = c_{\text{def}}(u^2 + v^2)\omega_0\alpha_c\,, \tag{3.50}$$

where

$$\alpha_c \simeq \frac{\pi}{N^2}\sum_q\left(-\frac{\partial\overline{n}_q}{\partial\epsilon_q}\right)\sum_{q'}|f_{q,q'}|^2\delta(\epsilon_{q'} - \epsilon_q)\,. \tag{3.51}$$

Let us evaluate (3.50) for $f_{q,q'} = f = \text{const.}$ and $\epsilon_q = (\hbar q)^2/2m_{\text{el}}$, where m_{el} is the mass of the conduction electron. Using the substitution

$$\frac{1}{N}\sum_q \to V_0\int\frac{d^3q}{(2\pi)^3} \tag{3.52}$$

where V_0 is the volume of the elementary cell, we obtain:

$$\frac{1}{N}\sum_{q'}|f_{q,q'}|^2\delta(\epsilon_{q'} - \epsilon_q) \simeq \frac{1}{\pi^2 2^{1/2}}\frac{V_0 m_{\text{el}}^{3/2}|f|^2\epsilon_q^{1/2}}{\hbar}\,. \tag{3.53}$$

Taking the derivative of $\overline{n}_q$ on ϵ_q

$$\frac{\partial \overline{n}_q}{\partial \epsilon_q} = \frac{1}{4 k_{\rm B} T \cosh^2(x - x_{\rm F})} \,, \tag{3.54}$$

where $x = \epsilon_q / 2 k_{\rm B} T$ and $x_{\rm F} = \epsilon_{\rm F} / 2 k_{\rm B} T$, we can calculate the sum on $\boldsymbol{q}$:

$$\frac{1}{N} \sum_q \left(-\frac{\partial \overline{n}_q}{\partial \epsilon_q} \right) \epsilon_q^{1/2} \simeq \frac{1}{\pi^2 2^{3/2}} \frac{V_0 m_{\rm el}^{3/2} k_{\rm B} T}{\hbar^3} \int_0^\infty \frac{x\, dx}{\cosh^2(x - x_{\rm F})} \,. \tag{3.55}$$

Changing the x variable as $x = y + x_{\rm F}$, one can estimate the integral in (3.55) as:

$$\int_{-x_{\rm F}}^{\infty} \frac{(y + x_{\rm F})\, dy}{\cosh^2 y} = 2(\ln 2 + x_{\rm F}) - \int_{x_{\rm F}}^{\infty} \frac{y\, dy}{\cosh^2 y} \,. \tag{3.56}$$

This integral for $x_{\rm F} \gg 1$ ($T \ll \epsilon_{\rm F}/2k_{\rm B} \sim 10^4$ K) approximately equals $2x_{\rm F}$. Combining all calculations together, we obtain

$$\eta_{\rm m-el} = \frac{c_{\rm def}}{4\pi^3} (u^2 + v^2) \omega_0 \frac{m_{\rm el}^3 V_0^2 |f|^2 \epsilon_{\rm F}}{\hbar^6} \,. \tag{3.57}$$

This relaxation rate is linear with the defect concentration $c_{\rm def}$ and frequency dependent: it contains the transformation terms

$$(u^2 + v^2) \omega_0 = \frac{\gamma (H_{x0} + H_{y0})}{2} \,. \tag{3.58}$$

Thus one has

$$\eta_{\rm m-el} \simeq \frac{c_{\rm def}}{8\pi^3} \gamma (H_{x0} + H_{y0}) \frac{m_{\rm el}^3 V_0^2 |f|^2 \epsilon_F}{\hbar^6} \,. \tag{3.59}$$

3.3 Relaxation via an Intermediate Damped Dynamic System

There is a whole class of relaxation mechanisms that cannot be analyzed as elementary processes. The energy loss in this case occurs via an intermediate damped dynamic system (see, for example, [10, 47, 50–54]). Here we shall consider the magnetization damping via the so-called slow-relaxing impurities (see Figure 3.2). In this mechanism the magnetization motion modulates the impurity splitting (levels in Figure 3.2). Thus the thermal equilibrium population of the energy levels varies, and transitions between the levels occur (arrows in Figure 3.2). There is a delay for these transitions due to a finite impurity relaxation time. This delay results in a magnetization oscillation energy loss.

Let us consider two-level impurities as effective spins $s_j = 1/2$. The anisotropic exchange Hamiltonian between the impurities and the neighboring host spins can

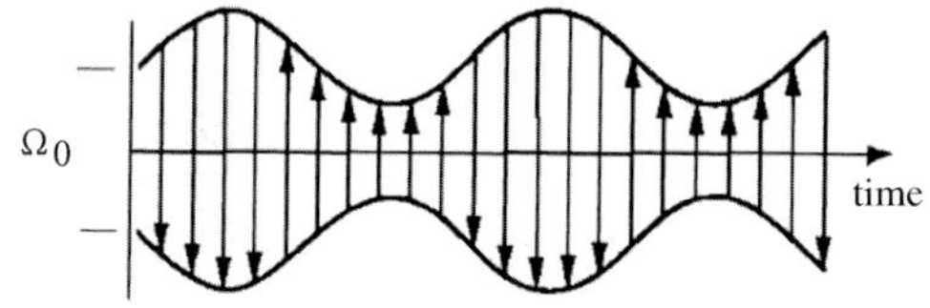

Figure 3.2 "Slow relaxation" mechanism of magnon relaxation by modulation of impurity levels.

be written in a general form:

$$\mathcal{H}_{\text{ex}} = \sum_{j,\nu} \sum_{c_j,c} B_{c_j,c}(\boldsymbol{R}_j, \boldsymbol{r}_\nu) s_{c_j}(\boldsymbol{R}_j) S_c(\boldsymbol{R}_j + \boldsymbol{r}_\nu) . \tag{3.60}$$

Here, $B_{c_j,c}(\boldsymbol{R}_j, \boldsymbol{r}_\nu)$ are the anisotropic exchange integrals, $c = x, y, z$; $c_j = x_j, y_j, z_j$ are the local principal axes for the jth impurity "spin" $s_j(\boldsymbol{R}_j) = \{s_{x_j}(\boldsymbol{R}_j), s_{y_j}(\boldsymbol{R}_j), s_{z_j}(\boldsymbol{R}_j)\}$ located at $\boldsymbol{R}_j$, and $\boldsymbol{S}(\boldsymbol{R}_j + \boldsymbol{r}_\nu) = \{S_x(\boldsymbol{R}_j + \boldsymbol{r}_\nu), S_y(\boldsymbol{R}_j + \boldsymbol{r}_\nu), S_z(\boldsymbol{R}_j + \boldsymbol{r}_\nu)\}$ are the components of the host spin located in the vicinity of jth impurity.

The Hamiltonian describing the impurity level modulation follows from (3.60):

$$\begin{aligned}\mathcal{H}_{\text{ex,slow}} = \frac{1}{\hbar} \sum_{j,\nu} \big[& B_{z_j,x}(\boldsymbol{R}_j, \boldsymbol{r}_\nu) S_x(\boldsymbol{R}_j + \boldsymbol{r}_\nu) \\ & + B_{z_j,y}(\boldsymbol{R}_j, \boldsymbol{r}_\nu) S_y(\boldsymbol{R}_j + \boldsymbol{r}_\nu) \big] s_{z_j}(\boldsymbol{R}_j) .\end{aligned} \tag{3.61}$$

In the case of only the coherent motion of the host spins we have $\boldsymbol{S}(\boldsymbol{R}_j + \boldsymbol{r}_\nu) = -(V_0/\hbar\gamma)\boldsymbol{M}$, where $V_0 = V/N$ is the volume of elementary cell. The magnetization components in terms of normal mode is expressed as

$$\begin{aligned} M_x &\simeq -\left(\frac{\hbar\gamma M_s}{2V}\right)^{1/2} \left(\frac{H_y}{H_x}\right)^{1/4} (b^\dagger + b) , \\ M_y &\simeq i\left(\frac{\hbar\gamma M_s}{2V}\right)^{1/2} \left(\frac{H_x}{H_y}\right)^{1/4} (b^\dagger - b) . \end{aligned} \tag{3.62}$$

Consider a two-level paramagnetic impurity with the energy

$$\mathcal{H}_{\text{imp},j} = \hbar \sum_j \left[\Omega_{0,j} + \delta\Omega_j(t) \right] n_j , \tag{3.63}$$

where $\Omega_{0,j}$ is the splitting frequency, $n_j = s_{z_j}(\boldsymbol{R}_j) + 1/2$ is the upper lever population and j is the impurity index. Applying (3.62) to (3.61), we can write the impurity level modulation as

$$\delta\Omega_j(t) = \Phi_j b(t) + \Phi_j^* b^\dagger(t) , \tag{3.64}$$

where

$$\Phi_j = \frac{V_0}{\hbar}\left(\frac{M_s}{\hbar\gamma V}\right)^{1/2}\sum_\nu\left[B_{z_j,x}(\boldsymbol{R}_j,\boldsymbol{r}_\nu)\left(\frac{H_y}{H_x}\right)^{1/4}\right.$$
$$\left.+iB_{z_j,y}(\boldsymbol{R}_j,\boldsymbol{r}_\nu)\left(\frac{H_x}{H_y}\right)^{1/4}\right]. \tag{3.65}$$

The kinetics of the impurity population is defined by the following equation:

$$\frac{dn_j}{dt} = -\Gamma_{\parallel,j}[n_j - n_T(\Omega_j)]\,. \tag{3.66}$$

Here $\Gamma_{\parallel,j}$ is the impurity relaxation rate and

$$n_T(\Omega_j) = \frac{1}{\exp(\hbar\Omega_j/k_B T)+1}$$

is the equilibrium population at frequency $\Omega_j = \Omega_{0,j} + \delta\Omega_j(t)$. Taking into account

$$n_j(t) = n_T(\Omega_{0,j}) + \delta n_j(t)\,,$$
$$n_T(\Omega_j) = n_T(\Omega_{0,j}) + \frac{\partial n_T}{\partial\Omega_{0,j}}\delta\Omega_j(t)\,, \tag{3.67}$$

we can solve (3.66) and obtain

$$\delta n_j(t) = \Gamma_{\parallel,j}\frac{\partial n_T}{\partial\Omega_{0,j}}\left[\frac{\Phi_j b(t)}{\Gamma_{\parallel,j}-i\omega_0}+\frac{\Phi_j^* b^\dagger(t)}{\Gamma_{\parallel,j}+i\omega_0}\right]. \tag{3.68}$$

The dynamic equation for the normal mode interacting with impurities is obtained by substituting (3.64) into (3.63). This yields an additional term to the basic mode Hamiltonian (3.41). The dynamic equation is now:

$$\frac{db}{dt} = -i\omega_0 b - \frac{i\partial(\mathcal{H}_{\text{imp},j}/\hbar)}{db*}$$
$$= -i\omega_0 b - i\sum_j \Phi_j^*\delta n_j(t)\,. \tag{3.69}$$

Here for simplicity, instead of using Bose operators, we use corresponding complex amplitudes b and b^*.

Substituting (3.68) into (3.67) and (3.69), we obtain the equation of damped harmonic oscillator:

$$\frac{db}{dt} + \eta_{\text{sr}} b = -i(\omega_0 + \Delta\omega_{\text{sr}})b\,. \tag{3.70}$$

Here the frequency shift is equal to

$$\Delta\omega_{\text{sr}} \simeq -\sum_j |\Phi_j|^2\left(-\frac{\partial n_T(\Omega_{0,j})}{\partial\Omega_{0,j}}\right)\frac{\Gamma_{\parallel,j}^2}{\Gamma_{\parallel,j}^2+\omega_0^2} \tag{3.71}$$

and the relaxation rate is

$$\eta_{\rm sr} \simeq \sum_j |\Phi_j|^2 \left(-\frac{\partial n_T(\Omega_{0,j})}{\partial \Omega_{0,j}}\right) \frac{\omega_0 \Gamma_{\parallel,j}}{\Gamma_{\parallel,j}^2 + \omega_0^2} , \tag{3.72}$$

where

$$|\Phi_j|^2 = \frac{M_{\rm s} V_0}{\hbar\gamma N} \left[\sum_\nu B_{z_j,x}(\boldsymbol{R}_j, \boldsymbol{r}_\nu)\right]^2 \left(\frac{H_y}{H_x}\right)^{1/2} + \left[\sum_\nu B_{z_j,y}(\boldsymbol{R}_j, \boldsymbol{r}_\nu)\right]^2 \left(\frac{H_x}{H_y}\right)^{1/2} . \tag{3.73}$$

To simplify (3.72), we assume that $\Omega_{0,j} = \Omega_0$ and $\Gamma_{\parallel,j} = \Gamma_\parallel$ for all impurities. Summing over j gives an average anisotropic exchange B:

$$\sum_j \left[\sum_\nu B_{z_j,x}(\boldsymbol{R}_j, \boldsymbol{r}_\nu)\right]^2 \simeq \sum_j \left[\sum_\nu B_{z_j,y}(\boldsymbol{R}_j, \boldsymbol{r}_\nu)\right]^2 = N_{\rm imp}(z_{\rm imp} B)^2 , \tag{3.74}$$

where $N_{\rm imp}$ is the total number of impurities in the sample and $z_{\rm imp}$ is the average number of magnetic neighbors for one impurity. Finally the relaxation rate (3.72) is:

$$\eta_{\rm sr} \simeq c_{\rm imp} S \frac{(z_{\rm imp} B)^2}{\hbar k_{\rm B} T} \frac{\exp(\hbar\Omega_0/k_{\rm B}T)}{[\exp(\hbar\Omega_0/k_{\rm B}T) + 1]^2} \frac{\gamma(H_x + H_y)\Gamma_\parallel}{\Gamma_\parallel^2 + \omega_0^2} , \tag{3.75}$$

where $c_{\rm imp} = N_{\rm imp}/N$ is the impurity concentration and $S = M_{\rm s} V_0/\hbar\gamma$ is the value of the host spin. This relaxation rate exhibits a definite temperature dependance ($\Gamma_\parallel$ also depends on T). As discussed in the previous section for magnon–electron scattering processes, the impurity relaxation rate also is frequency dependent, via the term $\gamma(H_x + H_y)$ (with identical dependence as in magnon–electron scattering).

3.4 Ferromagnetic Resonance Linewidth

FMR linewidth measurements have been widely utilized to determine the relaxation rate for low-level linear excitations (for example [56, 57]). In this section we will use the relaxation rates for microscopic mechanisms evaluated in previous sections to analyze a set of FMR data in soft thin films (NiFe, permalloy).

For small relaxation compared to the resonant frequency the linewidth is given by

$$\Delta\omega = 2\eta_0 \tag{3.76}$$

for all microscopic processes. Usually in FMR experiments, the field swept linewidth ΔH is measured. There is a simple relation between ΔH and frequency linewidth $\Delta\omega$, which is valid for the applied magnetic field parallel or perpendicular to the film plane:

$$\Delta H = \frac{\Delta\omega}{\partial\omega_0/\partial H_0} , \tag{3.77}$$

and then from (3.42)

$$\frac{\partial\omega_0}{\partial H_0} = \frac{\gamma^2(H_x + H_y)}{2\omega_0} . \tag{3.78}$$

Let us summarize the linewidths obtained previously. Using (3.59), (3.75), (3.77) and (3.78), we can obtain the field swept linewidths:

magnon–electron confluence process

$$\Delta H_{\text{m-el}} \simeq \frac{c_{\text{def}}}{2\pi^3}\frac{\omega_0}{\gamma}\frac{m^3 V_0^2 |f|^2 \epsilon_{\text{F}}}{\hbar^4} , \tag{3.79}$$

slow-relaxing impurity process

$$\Delta H_{\text{sr}} \simeq \frac{4 c_{\text{imp}} S}{\hbar\gamma}\frac{(z_{\text{imp}} B)^2}{k_B T}\frac{\exp(\hbar\Omega_0/k_{\text{B}} T)}{[\exp(\hbar\Omega_0/k_{\text{B}} T) + 1]^2}\frac{\omega_0 \Gamma_{\|}}{\Gamma_{\|}^2 + \omega_0^2} . \tag{3.80}$$

The temperature dependence of ΔH in permalloy thin films with the static magnetic field in the film plane has been measured by Patton and Wilts [58]. As seen in Figure 3.3, ΔH exhibits a strong frequency dependence with a maximum in the vicinity of $T_{\max} \approx 80$ K. The temperature maximum shifts to slightly higher temperatures with increasing FMR frequency. Such a nonmonotonic temperature dependence of ΔH is typical for slow-relaxing impurities (see [10, 58]). However, the slow-relaxing impurity process alone cannot describe the experiment [58]. One can assume that the magnon–electron confluence process plays a role of a relaxation rate "baseline". Thus for analysis we combine the slow-relaxing impurity and magnon–electron confluence processes together:

$$\Delta H = \Delta H_{\text{sr}}(T, \omega_0, c_{\text{imp}}) + \Delta H_{\text{m-el}}(\omega_0, c_{\text{def}}) . \tag{3.81}$$

According to (3.80), the temperature $T_{\max}$ of the linewidth maximum is given by

$$\Gamma_{\|}(T_{\max}) \simeq \omega_0 . \tag{3.82}$$

The impurity damping $\Gamma_{\|}$ is expected to increase monotonically with temperature [10]. Thus $T_{\max}$ will increase with increasing ω_0 in agreement with experiments. The solid lines in Figure 3.3 represent a theoretical fit with (3.81). We see that the temperature (and frequency) dependence (3.81) describe well the range from about the peak to about room temperature.

For the slow relaxation mechanism the fit gives

$$\Delta H_{\text{sr}}\,[\text{Oe}] \simeq \frac{1.9\times 10^4}{T\,[\text{K}]}\frac{\exp(100/T\,[\text{K}])}{[\exp(100/T\,[\text{K}]) + 1]^2}\frac{\omega_0/\Gamma_{\|}}{1 + (\omega_0/\Gamma_{\|})^2} , \tag{3.83}$$

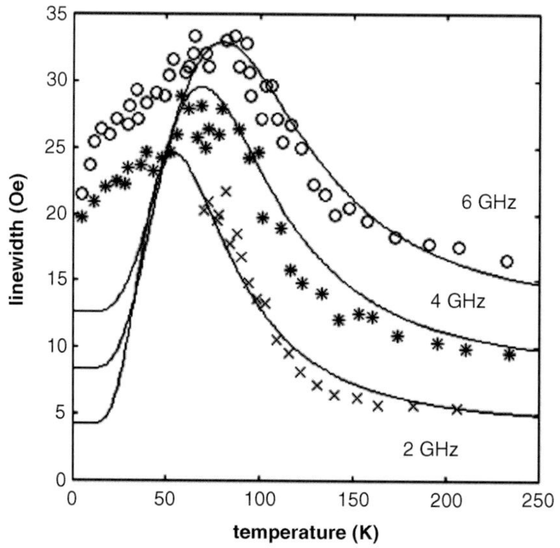

Figure 3.3 The linewidth temperature dependence for three FMR frequencies. Points are experimental data from [58]. Solid lines demonstrate theoretical fit.

where typical parameters for slow-relaxing impurity were used: $\hbar\Omega_0/k_B = 100$ K, $z_{imp}B/k_B = 25.3$ K, $S \simeq 1$ and $c_{imp} = 10^{-3}$. The spin–lattice relaxation rate variation with temperature was assumed to be $\Gamma_\parallel(T) = c_\parallel T^2$ with a fit of $c_\parallel/2\pi = 0.8 \times 10^{-3}$ GHz/K^2. The impurity interactions with conduction electrons are likely to be responsible for such a temperature dependence. The spin–lattice relaxation time at room temperature $1/\Gamma_\parallel(300\text{ K}) \simeq 2.2$ ps agrees with typical experimental data [10].

For the magnon–electron confluence process the fit gives

$$\Delta H_{m-el}\,[\text{Oe}] \simeq 2.1\left(\frac{\omega_0}{2\pi}\,[\text{GHz}]\right) . \tag{3.84}$$

The linear frequency dependence for ΔH_{m-el} was predicted in [51]. Substituting $m_{el} \simeq 10^{-30}$ kg, $V_0 \simeq 10^{-16}$ m^3 and $\gamma = 2.8$ MHz/Oe into (3.79) and comparing with (3.84), we obtain $c_{def}|\hbar f|^2\epsilon_F \simeq 3.5 \times 10^{-59}$ J^3. Taking defect concentration $c_{def} = 10^{-3}$ and Fermi energy $\epsilon_F = 3$ eV, one gets an estimate for the interaction amplitude $\hbar f \simeq 1.7$ eV, which is typical for ferromagnetic metals [60].

Patton *et al.* [59] have studied the frequency dependence of ΔH in thin permalloy films (17–48 nm) at room temperature (see Figure 3.4). They have found that the field linewidth has a linear frequency dependence $\Delta H \propto \omega_0$ for the case when the external magnetic field is parallel to the film plane (open circles in Figure 3.4). For the perpendicular case (black circles) the linear dependence is valid just for high FMR frequencies $\omega_0/2\pi > 8$ GHz and it is saturated at smaller frequencies. Recent data (R.D. McMichael, private communication) show similar results except that for the field perpendicular to the film plane the saturation region occurs only at extremely small frequencies ($\omega_0/2\pi < 2$ GHz). Essentially for a wide range of

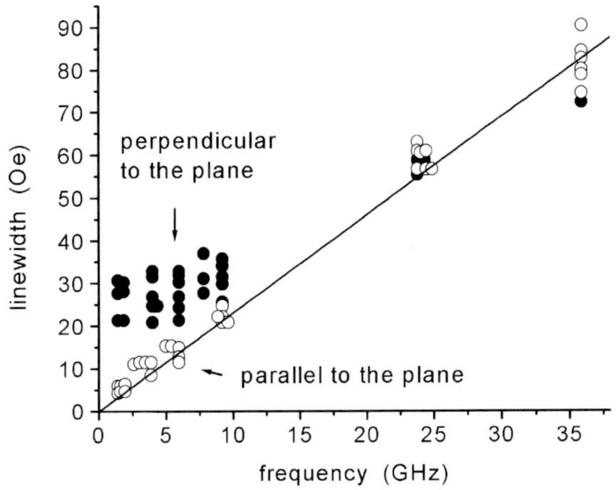

Figure 3.4 FMR linewidth versus frequency dependence at room temperature [59]. Solid straight line demonstrates theoretical fit.

frequencies the field swept linewidths are identical for the two field orientations and strictly linear with resonant frequency. It is clear that this linearity can be explained by the magnon–electron confluence process (3.84). At room temperature the impurity relaxation rate is $\Gamma_{\parallel}/2\pi \simeq 72\,\text{GHz}$ so that this mechanism yields a linewidth (3.83) that is also linear with frequency at least to about 50 GHz. Taking (3.84), (3.83) and (3.81) at $T = 300\,\text{K}$ we obtain:

$$\Delta H\,[\text{Oe}] \simeq 2.3\left(\frac{\omega_0}{2\pi}\,[\text{GHz}]\right) . \tag{3.85}$$

This linear dependence is shown in Figure 3.4 by a solid line and demonstrates good agreement with experiments.

We wish to point out that the thin films which were analyzed in [58] were surface oxidized (Carl Patton, private communication) and thus a large contribution of slow-relaxing impurities to the linewidth can be expected. Not all thin films will necessarily show analogous temperature dependence.

3.5 Magnons and Macroscopic Dynamic Equation

Both direct and indirect magnon interactions with a thermal bath yield damped harmonic oscillator dynamics in the form:

$$\frac{db}{dt} + \eta b = -i(\omega_0 + \Delta\omega)b . \tag{3.86}$$

Utilizing back transformation (3.43) and (3.34), from (3.86) (and its hermitian conjugate) we can derive linearized equations for the transverse magnetization components:

$$\frac{d}{dt}\begin{pmatrix} M_x \\ M_y \end{pmatrix} = \begin{pmatrix} -\eta & -\gamma H_y \\ \gamma H_x & -\eta \end{pmatrix}\begin{pmatrix} M_x \\ M_y \end{pmatrix} . \tag{3.87}$$

Our aim is to compare these equations with the linearized phenomenological equations.

3.5.1 Linearized Landau–Lifshitz Equation

The dynamics of magnetization $\boldsymbol{M}$ of a single-domain ferromagnetic sample is usually described by the Landau–Lifshitz equation [14]:

$$\frac{d\boldsymbol{M}}{dt} = -\gamma \boldsymbol{M} \times \boldsymbol{H}_{\text{eff}} - \frac{\alpha\gamma}{M_s}\boldsymbol{M} \times (\boldsymbol{M} \times \boldsymbol{H}_{\text{eff}}) , \tag{3.88}$$

where $\boldsymbol{H}_{\text{eff}}$ is the effective field and α is a dimensionless damping parameter.

The first term in (3.88) directly follows from microscopic equations and describes an averaged precession of a large number of spins. The second term in (3.88) was introduced phenomenologically from a simple geometric consideration to describe the magnetization damping with $|\boldsymbol{M}| = M_s$. Later Gilbert [62, 63] suggested that this damping term can be rewritten as a "dry friction" $\propto \boldsymbol{M} \times (d\boldsymbol{M}/dt)$ (see, also, [61]). Here we focus on the case of small damping ($\alpha \ll 1$) and neglect nonuniform magnetization motions.

Using (3.33), one can calculate the effective field $\boldsymbol{H}_{\text{eff}} = -\partial(\mathcal{E}/V)/\partial\boldsymbol{M}$ and write down the linearized equations (3.88) for the transverse magnetization components:

$$\frac{d}{dt}\begin{pmatrix} M_x \\ M_y \end{pmatrix} = \begin{pmatrix} -\alpha\gamma H_x & -\gamma H_y \\ \gamma H_x & -\alpha\gamma H_y \end{pmatrix}\begin{pmatrix} M_x \\ M_y \end{pmatrix} . \tag{3.89}$$

We see that the nondiagonal terms in (3.89) and (3.87), as expected, coincide with each other, respectively. The diagonal terms, responsible for relaxation, in general, are different ($H_x \neq H_y$). These damping terms are equal only in the special case of axial symmetry, when $H_x = H_y$. Thus we demonstrated that the conventional damping term in the Landau–Lifshitz equation applies only for the cases of high symmetry.

We can generalize (3.88) to the form [64]:

$$\frac{d\boldsymbol{M}}{dt} = -\gamma \boldsymbol{M} \times \boldsymbol{H}_{\text{eff}} - \gamma \frac{\boldsymbol{M}}{M_s} \times [\overleftrightarrow{\alpha} \cdot (\boldsymbol{M} \times \boldsymbol{H}_{\text{eff}})] . \tag{3.90}$$

Here a dimensionless damping tensor $\overleftrightarrow{\alpha}$ is introduced. Note that the new tensor damping conserves the length of the magnetization vector ($|\boldsymbol{M}| = M_s$) and gives a rate of change of the energy proportional to the square of the torque $\boldsymbol{M} \times \boldsymbol{H}_{\text{eff}}$.

The tensor $\overleftrightarrow{\alpha}$ should contain all necessary information about symmetry of the system. Such information is included in the expression for the energy of the system and can be expressed as a tensor $\partial^2(\mathcal{E}/V_0)/\partial \boldsymbol{M} \partial \boldsymbol{M}$. Thus, we can consider

$$\overleftrightarrow{\alpha} = \kappa \frac{\partial^2(\mathcal{E}/V)}{\partial \boldsymbol{M} \partial \boldsymbol{M}} = -\kappa \frac{\partial \boldsymbol{H}_{\text{eff}}}{\partial \boldsymbol{M}} , \tag{3.91}$$

where κ is a dimensionless parameter. Substituting the expression (3.33) for the energy $\mathcal{E}$ into (3.91), we obtain the damping tensor in the vicinity of equilibrium:

$$\overleftrightarrow{\alpha} = \kappa \begin{pmatrix} H_x/M_s & 0 & 0 \\ 0 & H_y/M_s & 0 \\ 0 & 0 & 0 \end{pmatrix} . \tag{3.92}$$

Equation (3.90) with the tensor $\overleftrightarrow{\alpha}$ (3.92) must be consistent with (3.87) for small magnetization oscillations. Linearizing (3.90) and comparing with (3.87), we obtain

$$\kappa = \frac{\eta M_s}{\gamma H_x H_y} . \tag{3.93}$$

Thus we see that the damping tensor appears from a general form of interaction of the normal modes of the magnetic system with a thermal bath. The tensor is scaled by only one magnon damping parameter η. This leads to the identical relaxation of transverse magnetization components, as in the Bloch–Bloembergen equations.

A dynamic equation for large magnetization motions is derived in Appendix D.

3.6
Relaxation of Coupled Oscillations

In this section we analyze the relaxation in the system of coupled oscillations. In the previous chapter we have already considered two examples of coupled oscillations in magnetic systems: nuclear magnons and quasi phonons which appear as the result of linear interaction of quasi ferromagnons with nuclear spin deviations and elastic vibrations of the lattice, respectively.

The frequencies of normal modes of the coupled oscillations can differ noticeably from the frequencies of original "pure" modes. The change in spectra also leads to the the change of relaxation. For example, both coupled oscillations become damped if even just one of the constituent pure modes is damped. In this section we consider the problem of finding the linewidths of two coupled oscillations. The most systematic method of solving this problem is to seek the line shapes of the new normal modes of the system with allowance for the coupling by proceeding from the known line shapes of the original pure modes. Recall that the shape of the resonance line $f_1(\Omega)$ is the normalized weight function, according to which the statistical-average parameters of the oscillation, the moments $\mathcal{M}_n$ are determined [12]:

$$\mathcal{M}_n = \int_{-\infty}^{\infty} (\Omega - \omega_1)^n f_1(\Omega) d\Omega , \tag{3.94}$$

where ω_1 is the eigenfrequency of the oscillator and $n = 1, 2, \ldots$ For example, for a harmonic oscillator with no damping

$$f(\Omega) = \delta(\Omega - \omega)\,, \quad \mathcal{M}_n = 0\,. \tag{3.95}$$

If an interaction is turned on between two oscillators with known eigenfrequencies ω_1, and ω_2, and line shapes $f_1(\Omega)$ and $f_2(\Omega)$, the line shape of any new normal mode of this coupled system can be represented in the form

$$\tilde{f}(\Omega) = \int\limits_{-\infty}^{\infty}\int\limits_{-\infty}^{\infty} \delta[\Omega - \tilde{\omega}(\Omega_1, \Omega_2)] f_1(\Omega_1) f_2(\Omega_2) d\Omega_1 d\Omega_2\,, \tag{3.96}$$

where $\tilde{\omega}(\omega_1, \omega_2)$ is the frequency of the corresponding normal mode as obtained from the characteristic equation. This expression reflects the fact that the line shape of the normal mode is formed with allowance for the interactions between the different spectral components of the original oscillations, taken with the corresponding weight factors. From (3.96) one can easily obtain the formula for the moments of the new normal mode:

$$\tilde{\mathcal{M}}_n = \int\limits_{-\infty}^{\infty}\int\limits_{-\infty}^{\infty} [\tilde{\omega}(\Omega_1, \Omega_2) - \tilde{\omega}(\omega_1, \omega_2)]^n f_1(\Omega_1) f_2(\Omega_2) d\Omega_1 d\Omega_2\,. \tag{3.97}$$

Expressions (3.96) and (3.97) give a general solution of the problem of finding the linewidths of the normal modes of two coupled oscillations [65].

Let us estimate the damping as

$$\tilde{\eta} \simeq \left(\tilde{M}_2\right)^{1/2}\,, \tag{3.98}$$

where the second moment can be calculated in the following way:

$$\begin{aligned} \tilde{\mathcal{M}}_2 &= \int\limits_{-\infty}^{\infty}\int\limits_{-\infty}^{\infty} \left[\tilde{\omega}(\Omega_1, \Omega_2) - \tilde{\omega}(\omega_1, \omega_2)\right]^2 f_1(\Omega_1) f_2(\Omega_2) d\Omega_1 d\Omega_2 \\ &\simeq \int\limits_{-\infty}^{\infty}\int\limits_{-\infty}^{\infty} \left[\frac{\partial\tilde{\omega}}{\partial\Omega_1}(\Omega_1 - \omega_1) + \frac{\partial\tilde{\omega}}{\partial\Omega_2}(\Omega_2 - \omega_2)\right]^2 f_1 f_2 d\Omega_1 d\Omega_2 \\ &\simeq \left[\frac{\partial\tilde{\omega}(\omega_1, \omega_2)}{\partial\omega_1}\right]^2 \mathcal{M}_2^{(1)} + \left[\frac{\partial\tilde{\omega}(\omega_1, \omega_2)}{\partial\omega_2}\right]^2 \mathcal{M}_2^{(2)}\,. \end{aligned} \tag{3.99}$$

Here we have assumed that the line shapes of the initial pure oscillations were symmetric (the first moments are equal to zero) and the frequencies

$$|\omega_1 - \omega_2| \gg \sqrt{\mathcal{M}_2^{(1)}}, \sqrt{\mathcal{M}_2^{(2)}}\,,$$

where $\mathcal{M}_2^{(1)}$ and $\mathcal{M}_2^{(2)}$ are the second moments of the first and second pure oscillators, respectively. If the first term in (3.99) is much less than the second one, we can write

$$\tilde{\eta}_2 \simeq \frac{\partial \tilde{\omega}_2(\omega_1, \omega_2)}{\partial \omega_1} \eta_1 \,. \qquad (3.100)$$

This formula has a simple physical meaning that the damping η_1 of the initial pure oscillation of the frequency ω_1 is "transferred" to the damping $\tilde{\eta}_2$ of a new normal mode of the frequency $\tilde{\omega}_2$.

3.6.1
Example 1: Nuclear Magnons

As an example, we shall consider nuclear magnons in an antiferromagnet. Nuclear magnons, which appear as a result of coupled oscillations of quasi-ferromagnetic magnons ($\omega_{\mathrm{f}k}$) and Larmor nuclear spins precession (ω_{n}), represent a good candidate to check the validity of the "transferred" damping. The formula (3.100) for nuclear magnons can be represented as

$$\eta_{\mathrm{n}k} \simeq \frac{\partial \omega_{\mathrm{n}k}}{\partial \omega_{\mathrm{f}k}} \eta_{\mathrm{f}k} \,, \qquad (3.101)$$

where

$$\frac{\partial \omega_{\mathrm{n}k}}{\partial \omega_{\mathrm{f}k}} = \frac{(1-\xi^2)^{1/2}}{\xi} \frac{\omega_{\mathrm{n}}}{\gamma H_\Delta} \,, \qquad \xi = \frac{\omega_{\mathrm{n}k}}{\omega_{\mathrm{n}}} \,.$$

Here $\eta_{\mathrm{f}k}$ is the relaxation rate of the pure quasi ferromagnons. So far as the difference between the "old", pure quasi ferromagnons and "new" quasi ferromagnons that are decoupled from the nuclear spin precession is negligibly small, we can neglect the difference between the relaxation rate $\eta_{\mathrm{f}k}$ in (3.101) and the relaxation rate of new quasi ferromagnons $\tilde{\eta}_{\mathrm{f}k}$.

The most direct check to confirm the theory is to demonstrate that a specific feature in the damping of quasi ferromagnons is transferred to the damping of nuclear magnons. Experimentally it was demonstrated that the relaxation rate of quasi ferromagnetic magnons in $CsMnF_3$ exhibits a so-called magnon–phonon peak ($\sim$ 30% above the expected value of linewidth) that corresponds to the magnetoelastic crossover effects when the quasi ferromagnon spectrum crosses the transverse phonon spectrum [66]. This crossover is shown schematically in Figure 3.5b. According to (3.101), we can expect the relaxation peak of the nuclear magnon relaxation at the point (see dashed arrow), where no anomaly is expected. It should be emphasized that the crossover of the nuclear magnons spectrum with the phonons occurs at very small values of wave vector and, consequently, cannot mask the expected effect.

Figure 3.5 shows the k dependence of the nuclear magnon relaxation rate in $CsMnF_3$ ($\omega_{\mathrm{n}}/2\pi = 666\,\mathrm{MHz}$, $\gamma H_\Delta/2\pi = 7.08/\sqrt{T\,[\mathrm{K}]}\,\mathrm{GHz}$). In addition to a rapid growth of the relaxation at large k (small H) due to scattering of nuclear magnons by domain walls, we clearly see a relaxation peak corresponding to

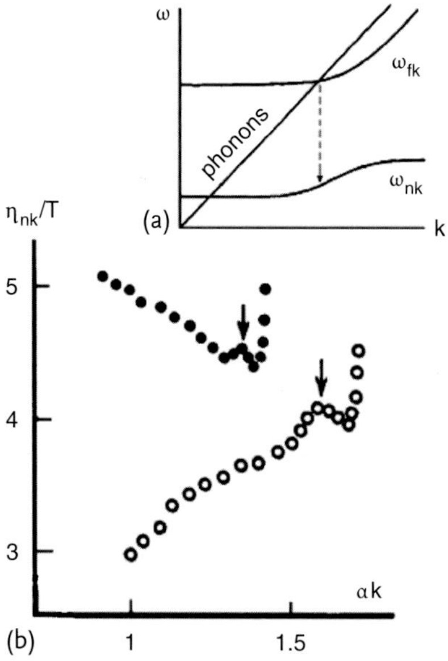

Figure 3.5 (a) Behavior of the nuclear magnon relaxation in $CsMnF_3$, in the coordinates η_{nk}/T [2π kHz/K] versus αk [kOe]($\alpha = 0.95 \times 10^{-5}$ kOe cm (in SI: 1 kOe = 79.58 kA/m) at two temperatures: $T = 4.23$ K (filled dots) and $T = 3.03$ K (open dots). The nuclear magnon frequency is $\omega_{nk}/2\pi = 511$ MHz. The arrows indicate the positions of the transverse magnon–phonon peak transferred from the quasi ferromagnon relaxation. (b) Spectra of quasi ferromagnons (ω_{fk}), nuclear magnons (ω_{nk}) and phonons. The dashed line connects the points where the quasi ferromagnon peak relaxation is transferred to nuclear magnon relaxation.

the magnon–phonon spectra crossover. The relative increase (~ 5%) in the relaxation rate at the peak is in a good agreement with the result expected from formula (3.101). The transferred damping mechanism also explains a number of the other important experimental details [65].

3.6.2
Example 2: Magnetoelastic Oscillations

Let us consider another example of transferred relaxation in a system of coupled oscillations, the case of magnetoelastic waves (and their quanta, quasi phonons).

The relaxation of magnetoelastic waves is a complex process which is determined by interactions of elastic and magnetic components of a wave with various subsystems of a crystal. We assume that the damping of quasi phonons consists of two parts:

$$\eta_{\mathrm{ph},k} = \eta_{\mathrm{ph0},k} + \frac{\partial \Omega_k}{\partial \omega_{\mathrm{f}k}} \eta_{\mathrm{f}k} \,. \tag{3.102}$$

This formula takes into account both the direct interaction of new normal modes with a thermal bath ($\eta_{\mathrm{ph0},k}$) and the indirect damping transferred from the initial "pure" quasi-ferromagnetic oscillations. Here

$$\Omega_k = v_e k \sqrt{1 - \delta_k} \tag{3.103}$$

is the spectrum of quasi phonons, where $\delta_k = (\gamma H_{\Delta me} \zeta / \omega_{fk})^2$, and

$$\omega_{fk} = \gamma \sqrt{H(H + H_D) + H_{\Delta me}^2 + (\alpha k)^2} \tag{3.104}$$

is the spectrum of quasi ferromagnons and η_{fk} is their damping.

We shall follow the experiment from [67], where the inverse lifetime $\tau_{ph}^{-1} = \eta_{ph}$ of the magnetoelastic oscillation of the sample was studied. The "easy plane" magnetic anisotropy antiferromagnet $FeBO_3$ (Neel temperature, $T_N = 348$ K) has the Dzyaloshinskii field $H_D = 100$ kOe (50 kOe) (in SI: $H_D = 79.58 \times 10^5$ A/m (39.79×10^5 A/m)); the magnetoelastic gap in the spin wave spectrum, $\gamma H_{\Delta me} = 5.32$ GHz (1.96 GHz); and the constant of inhomogeneous exchange $\alpha = 0.08$ Oe cm (in SI: $\alpha = 6.36 \times 10^{-2}$ A).

Sound velocity v_e and coefficient ζ, which describes the linear interaction of quasi ferromagnons and phonons, depend on the direction of wave vector and polarization of quasi phonons. The work in [67] studied the magnetoelastic excitations with wave vector oriented along the main axis (C_3) of the crystal and with the polarization parallel to the static magnetic field $\boldsymbol{H}$. The most strongly coupled with the magnetic subsystem quasi phonons of this polarization have velocity of $v_e \simeq 4.8 \times 10^3$ m/s and $\gamma H_{\Delta me} \zeta \simeq 4.76$ GHz (1.4 GHz). All above parameters given in parentheses were measured at $T = 300$ K. These parameters are different from those at low temperatures $T \ll T_N$.

The sample was a plate with a characteristic linear dimension $\simeq 4$ mm and thickness $d = 0.4$ mm. The developed planes of the plate coincided with the antiferromagnetic easy plane. The eigenfrequencies of oscillations of the plate sample can be calculated if we substitute in (3.103) the value of wave vector $k = (\pi/d)n$, where $n = 1, 2 \ldots$ is the number of magnetoelastic modes. Experimentally studied the eigenmode with the minimal wave vector π/d (minimal eigenfrequency), is most strongly coupled to the alternating magnetic field. Taking into account (3.103), we can rewrite (3.102) in the form:

$$\tau_{ph}^{-1} = \tau_{ph0}^{-1} + \frac{\Omega_e}{\omega_{f0}} \frac{\delta_0}{1 - \delta_0} \tau_m^{-1} , \tag{3.105}$$

where $\tau_m^{-1} = \eta_{f0}$ is the inverse lifetime of the quasi ferromagnons with small k.

Figure 3.6 shows the dependencies of the inverse lifetime of magnetoelastic oscillation on the static magnetic field. A strong field dependence of τ_{ph}^{-1} indicates that the damping of magnetoelastic waves in all temperature ranges is due to a fast-relaxing magnetic subsystem in the region of small fields. The solid lines in Figure 3.6 are drawn by (3.105) with $\tau_{ph0}^{-1} = 10^3\,s^{-1}$ and $\tau_m^{-1} = 0.2 \times 10^9 s^{-1}$ at $T = 85$ K, and $\tau_{ph0}^{-1} = 15 \times 10^3\,s^{-1}$ and $\tau_m^{-1} = 2 \times 10^9\,s^{-1}$ at $T = 300$ K, respectively. In this calculation the following experimental values of δ_0 were used: $\delta_0 = 14/(H + 100)$ at $T = 85$ K and $\delta_0 = 3.1/(H + 90)$ at $T = 300$ K, respectively (here the magnetic field H is given in Oe).

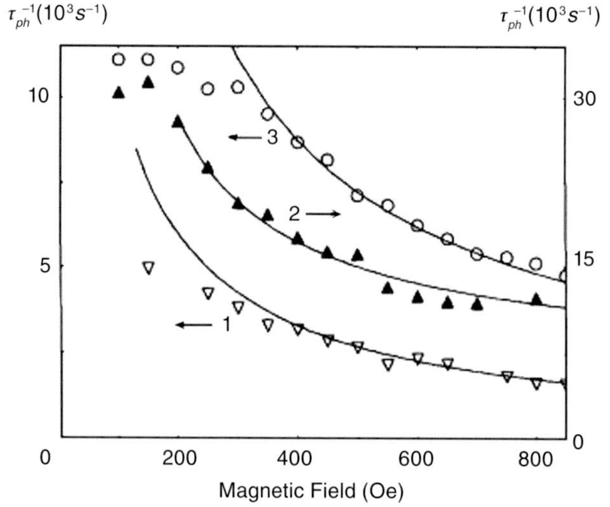

Figure 3.6 The magnetic field (100 Oe = 7958 A/m) dependence of the inverse lifetime of magnetoelastic oscillation at temperatures 85 K (curves 1 and 2) and 300 K (curve 3). Curve 1 corresponds to measurements on a free sample. Curve 2 was obtained on the sample in which one of the developed planes was covered by a thin film of glue. Solid lines correspond to the theoretical calculations (3.105).

An additional testing of expression (3.105) were carried out by means of an artificial change of damping in the magnetic subsystem. As it was reported in [68], the damping of quasi ferromagnons strongly increases if there is an increase of nonhomogeneous strains due to glueing of the sample. To create such nonhomogeneous strains, one of the developed planes of the sample was covered by a thin glue film at room temperature. In the control experiment, the relaxation rate of quasi ferromagnons of the frequency 18 GHz excited by a parallel pumping at $T = 4.2$ K increased five times. The effect of the glue film on the relaxation rate of magnetoelastic oscillation was also observed at $T = 85$ K. The result is given in Figure 3.6 (curve 2). We see that the relaxation rate is about three times greater, but the magnetic field dependence is approximately the same as in the case without glue film (curve 1). This demonstrated that the observed increase of magnetoelastic mode damping is directly associated with the increase of relaxation rate of quasi ferromagnetic oscillations. "Pure elastic" losses (τ_{ph0}^{-1}) at the same time were negligibly small as in the case of curve 1.

Thus, the experiment confirms the model of damping transfer in a system of coupled oscillations (see, also, [69]). An artificial increase of damping of quasi ferromagnetic oscillations by the nonhomogeneous strains in the sample also agrees with a corresponding increase of magnetoelastic mode damping, which testifies in favor of the considered model.

3.7 Discussion

a) Why do magnons have finite lifetimes?
b) Why is the system of harmonic oscillators a good model of a thermal bath?
c) Discuss the role of a system of conduction electrons as a thermal bath.
d) Discuss the role of crystalline defects for the magnon relaxation. Can you control magnetic relaxation using defects and impurities?
e) Discuss the role of sample surface as a defect and its role in magnon relaxation.
f) Which way is more preferable (why and when) for computer simulations of the magnetic dynamics of the magnetic material: (i) to represent the material as a system of individual magnetic "particles" interacting with each other and solve Landau–Lifshitz–Gilbert equations for these "particles" or, (ii) to find collective modes (magnons) in the system and represent magnetic dynamics in terms of dynamics of magnons.

4
Microwave Pumping of Magnons

Magnons can be excited by the external microwave magnetic field applied to the magneto-ordered system. Depending on the geometry, frequency and power, the growth of magnon population can be caused by linear or nonlinear processes. The simplest example of a linear process is the ferromagnetic resonance when the uniform magnetization precession (magnons with $\boldsymbol{k} = 0$) is excited by the transverse alternating magnetic field with frequency ω_0. Nonlinear processes open the possibility of exciting magnons with $\boldsymbol{k} \neq 0$ via parametric resonance conditions.

One of the most frequently used and powerful techniques is the method of so-called parallel pumping when the microwave field is parallel to the DC magnetic field. In this case the decay parametric process occurs: a microwave photon creates a pair of magnons with the oppositely directed wave vectors and the energies $\hbar\omega_{\boldsymbol{k}} = \hbar\omega_{-\boldsymbol{k}} = \hbar\omega_{\mathrm{p}}/2$. The photon wave vector in this case is negligibly small and can be considered as zero. This process has a pumping field threshold above which the number of magnon pairs grows exponentially.

Nonlinearities of the system are responsible for the restriction of this growth.

Principal ideas of the theory of parametric resonance excitation of spin waves (and magnons) were proposed by Suhl [70] and Schlömann [71]. Zakharov, L'vov and Starobinets [72] (see also [73]) have developed a basic theoretical model known as S theory, for the above process ($\omega_{\mathrm{p}} = \omega_{\boldsymbol{k}} + \omega_{-\boldsymbol{k}}$) of parametric spin-wave excitation in which two independent mechanisms for the flow stabilization can be taken into account: (i) the so-called phase mechanism due to four-magnon interactions, wherein the forced oscillations of the magnetic medium (parametric pair) deviate in phase from the pumping field, and (ii) the positive nonlinear damping, wherein the relaxation rate of the excited waves increases with the increase of their amplitudes.

The S theory has considered the magnetic system as it is placed in an alternating magnetic field $h \cos \omega_{\mathrm{p}} t$. High amplitudes of the microwave magnetic field are usually obtained with the help of microwave resonators. The resonator, as it was expected earlier (see for example [10, 73]), does not play a specific role in the process of parametric spin wave excitation. It was assumed to cause only an additional linear microwave absorption which can be easily taken into account in order to calculate the microwave power absorbed by the excited magnetic system of the sample (and therefore, nonlinear magnetic susceptibility, number of parametric magnons

Nonequilibrium Magnons, First Edition. Vladimir L. Safonov.

and so on). However, it was shown [74, 75] that there is a principal nonlinear effect of the microwave resonator on the process of parallel pumping of magnons. In this chapter we shall consider the process of parametric excitation of magnons taking into account specific dynamics of the resonator cavity mode.

To derive dynamic equations, one can utilize the master equation approach we considered in the previous chapter. It is valid also for the description of the system with time dependent Hamiltonian $\mathcal{H} + \delta\mathcal{H}(t)$ if the energy variation is small compared to the energy of the system. In this case the equation for the mean value of an operator $\mathcal{O}$ becomes

$$\begin{aligned}\frac{d}{dt}\langle\mathcal{O}\rangle &= -\frac{i}{\hbar}\langle[\mathcal{O}, \mathcal{H}_{\rm d} + \delta\mathcal{H}(t)]\rangle \\ &\quad - \int_0^\infty du \sum_{l,j} \{\langle F_l(u) F_j(0)\rangle_{\rm b}\langle[\mathcal{O}, Q_l] Q_j(-u)\rangle \\ &\quad - \langle F_j(0) F_l(u)\rangle_{\rm b}\langle Q_j(-u)[\mathcal{O}, Q_l]\rangle\} \,. \end{aligned} \tag{4.1}$$

So far as in this chapter we are going to work with the parametrically excited magnon states of the dynamical system, we can use classical canonical variables. For example, for the creation and annihilation operators of a magnon the (4.1) is reduced to

$$\begin{aligned}\left(\frac{d}{dt} + \eta_k\right) b_k &= -i\frac{\partial\left[\mathcal{H} + \delta\mathcal{H}(t)\right]/\hbar}{\partial b_k^*} \,, \\ \left(\frac{d}{dt} + \eta_k\right) b_k^* &= i\frac{\partial\left[\mathcal{H} + \delta\mathcal{H}(t)\right]/\hbar}{\partial b_k} \,. \end{aligned} \tag{4.2}$$

Note that these equations are completely analogous to Hamilton's equations for the complex spin-wave amplitudes supplemented by *phenomenological* damping parameters (see for example [4, 72, 73]). All magnons outside the parametric resonance conditions are assumed to be a part of a thermal bath.

4.1
Linear Theory

Let us consider the Hamiltonian of the magnons system in the form

$$\mathcal{H} = \mathcal{H}_0 + \mathcal{H}_t \,, \tag{4.3}$$

where

$$\frac{\mathcal{H}_0}{\hbar} = \sum_q \omega_q b_q^* b_q \tag{4.4}$$

describes the energy of an ideal magnon gas, and $\mathcal{H}_t = \delta\mathcal{H}(t)$ is the time-dependent Hamiltonian responsible for the excitation of magnons.

4.1.1 Ferromagnetic Resonance

The most simple time-dependent Hamiltonian,

$$\frac{\mathcal{H}_t}{\hbar} = f(e^{-i\omega t} b_0^* + e^{i\omega t} b_0) \tag{4.5}$$

describes linear excitation of the uniform magnetization precession, the ferromagnetic resonance (see Figure 4.1). The dynamic equation

$$\left(\frac{d}{dt} + \eta_0\right) b_0 = -i\omega_0 b_0 - i f e^{-i\omega t}, \tag{4.6}$$

has an almost obvious solution

$$b_0 = \frac{i f e^{-i\omega t}}{i(\omega_0 - \omega) + \eta_0}, \tag{4.7}$$

which describes the Lorentzian line shape of absorption.

4.1.2 Threshold of Parametric Resonance

In order to excite magnons with $\boldsymbol{k} \neq 0$ by the external field with $\boldsymbol{k} = 0$ we have to consider conditions of parametric resonance. The most frequently used time-dependent Hamiltonian

$$\frac{\mathcal{H}_t}{\hbar} = \frac{h V_k}{2}\left[b_k b_{-k} \exp(i\omega_p t) + b_k^* b_{-k}^* \exp(-i\omega_p t)\right] \tag{4.8}$$

describes the classical microwave pumping field $h(t) \propto \exp(-i\omega_p t)$, which excites magnon pair (see Figure 4.1) via the coupling coefficient V_k. The term ω_p is the frequency of the pump field.

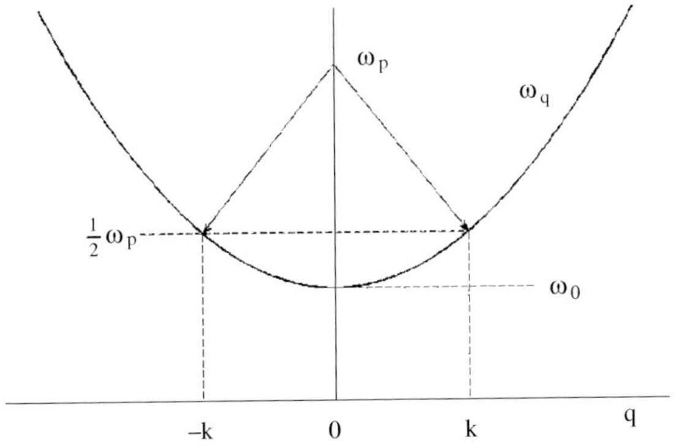

Figure 4.1 Schematic picture representing conditions of linear (ω_0) and parametric ($\omega_k = \omega_{-k} = \omega_p/2$) resonances of magnons.

Substituting (4.4) and (4.8) into (4.2), one obtains the dynamic equations for canonical complex amplitudes

$$\left(\frac{d}{dt} + \eta_k\right) b_k = -i\omega_k b_k - ihV_k e^{-i\omega_p t} b^*_{-k} ,$$
$$\left(\frac{d}{dt} + \eta_k\right) b^*_{-k} = i\omega_k b^*_{-k} + ihV_k e^{i\omega_p t} b_k . \tag{4.9}$$

We see that the variables b_k and b^*_{-k} are coupled to each other. The same is valid for b_{-k} and b^*_k. The meaning of this coupling is that the pair of magnons with $\boldsymbol{k}$ and $-\boldsymbol{k}$ is excited simultaneously by the pumping field. Let us introduce the so-called slow variables c_k and c^*_{-k} by:

$$b_k = c_k \exp\left(-\frac{i\omega_p t}{2}\right) ,$$
$$b^*_{-k} = c^*_{-k} \exp\left(\frac{i\omega_p t}{2}\right) . \tag{4.10}$$

Then, (4.9) can be rewritten as

$$\left[i\left(\frac{d}{dt} + \eta_k\right) + \left(\frac{\omega_p}{2} - \omega_k\right)\right] c_k - hV_k c^*_{-k} = 0 ,$$
$$hV_k c_k + \left[i\left(\frac{d}{dt} + \eta_k\right) - \left(\frac{\omega_p}{2} - \omega_k\right)\right] c^*_{-k} = 0 . \tag{4.11}$$

This system of linear differential equations can be transformed to an algebraic form by substitution

$$c_k(t) = c_k(0) e^{\lambda t} ,$$
$$c^*_{-k}(t) = c^*_{-k}(0) e^{\lambda t} , \tag{4.12}$$

which gives

$$\begin{pmatrix} i(\lambda + \eta_k) + (\omega_p/2 - \omega_k) & -hV_k \\ hV_k & i(\lambda + \gamma_k) - (\omega_p/2 - \omega_k) \end{pmatrix} \begin{pmatrix} c_k(0) \\ c^*_{-k}(0) \end{pmatrix} = 0 . \tag{4.13}$$

This equation has a solution if the determinant is equal to zero:

$$\det \begin{pmatrix} i(\lambda + \eta_k) + (\omega_p/2 - \omega_k) & -hV_k \\ hV_k & i(\lambda + \gamma_k) - (\omega_p/2 - \omega_k) \end{pmatrix} = 0 ,$$

from which we obtain

$$(\lambda + \eta_k)^2 + \left(\frac{\omega_p}{2} - \omega_k\right)^2 - (hV_k)^2 = 0 \tag{4.14}$$

and

$$\lambda = -\eta_k \pm \sqrt{(h V_k)^2 - \left(\frac{\omega_p}{2} - \omega_k\right)^2} . \tag{4.15}$$

One can see that the condition of instability

$$\mathrm{Re}(\lambda) = 0 \tag{4.16}$$

first is valid for magnons with the resonance condition

$$\omega_k = \omega_{-k} = \frac{\omega_p}{2} . \tag{4.17}$$

The critical (threshold) amplitude for these magnons is

$$h_c = \frac{\eta_k}{V_k} . \tag{4.18}$$

Actually, the parametric instability begins for the pair with minimal h_c. The condition (4.18) in the form $h_c V_k = \eta_k$ simply means that "the pumping equals damping".

There is an infinite exponential growth of spin-wave complex amplitudes with $\boldsymbol{k}$ and $-\boldsymbol{k}$ wave vectors in the framework of linear theory above the threshold $h > h_c$.

$$c_k(t), c_{-k}(t) \propto \exp\left[\left(\frac{h}{h_c} - 1\right)\eta_k t\right] \tag{4.19}$$

Later we shall use the term "parametric pair" [76] to denote $\boldsymbol{k}$ and $-\boldsymbol{k}$ simultaneous excitations. Restrictions of this growth appear if we take into account the interactions of magnons or their nonlinear relaxation.

4.2 Parametric Resonance in a Resonator Cavity

Now we shall consider the magnetic system which is placed in a microwave resonator which is a harmonic oscillator

$$\frac{\mathcal{H}_R}{\hbar} = \omega_R R^* R . \tag{4.20}$$

Here R^* and R are the complex canonical variables of the resonator mode with the frequency ω_R.

The Hamiltonian of magnon pumping terms via resonator mode can be written as [78]:

$$\frac{\mathcal{H}_t}{\hbar} = F\left[R^* \exp(-i\omega_p t) + R \exp(i\omega_p t)\right] + \frac{i G_k}{2}\left(R b_k^* b_{-k}^* - R^* b_k b_{-k}\right) , \tag{4.21}$$

where G_k is the coupling parameter between the resonator mode and magnon pair; the term F is the amplitude of a pump field.

The equations of motion for complex amplitudes R and b_k have the form

$$i\left(\frac{d}{dt}+\Gamma\right)R=\frac{\partial(\mathcal{H}/\hbar)}{\partial R^*}\,, \tag{4.22}$$

$$i\left(\frac{d}{dt}+\eta_k\right)b_k=\frac{\partial(\mathcal{H}/\hbar)}{\partial b_k^*}\,, \tag{4.23}$$

where Γ and η_k are the damping parameters and $\mathcal{H}=\mathcal{H}_0+\mathcal{H}_R+\mathcal{H}_t$. In an explicit form one obtains

$$i\left(\frac{d}{dt}+\Gamma\right)R=\omega_R R+F\exp(-i\omega_p t)-\frac{iG_k}{2}b_k b_{-k}\,, \tag{4.24}$$

$$i\left(\frac{d}{dt}+\eta_k\right)b_k=\omega_k b_k+\frac{iG_k}{2}Rb_{-k}^*\,. \tag{4.25}$$

Introducing slow variables

$$\begin{aligned} R&=\tilde{R}\exp(-i\omega_p t)\,,\\ b_k&=c_k\exp(-i\omega_p t/2)\,, \end{aligned} \tag{4.26}$$

we can write the equations of motion for complex amplitudes as

$$\left[i\left(\frac{d}{dt}+\Gamma\right)+(\omega_p-\omega_R)\right]\tilde{R}=F-\frac{iG_k}{2}c_k c_{-k}\,, \tag{4.27}$$

and

$$\begin{aligned} &\left[i\left(\frac{d}{dt}+\eta_k\right)+\left(\frac{\omega_p}{2}-\omega_k\right)\right]c_k-\frac{iG_k}{2}\tilde{R}c_{-k}^*=0\,,\\ &-\frac{iG_k}{2}\tilde{R}^*c_k+\left[i\left(\frac{d}{dt}+\eta_k\right)-\left(\frac{\omega_p}{2}-\omega_k\right)\right]c_{-k}^*=0\,. \end{aligned} \tag{4.28}$$

One can see that a correspondence of (4.28) and (4.11) can be obtained if $\tilde{R}=F/i\Gamma$ and we put $hV_k=iG_k\tilde{R}/2=-iG_k\tilde{R}^*/2=G_kF/2\Gamma$. In this case $\Gamma\gg\eta_k$, $\omega_p=\omega_R$ and the last term in (4.27) is neglected, which is valid below the threshold of parametric instability. The nonlinearity $-i(G_k/2)c_k c_{-k}$ becomes important above the threshold field and brings a restriction of the exponential growth of parametric magnons.

Let us multiply expressions (4.28) and corresponding expressions for c_{-k} and c_k^* by c_k^*, c_{-k}^*, c_k, c_{-k} and combine equations for the following variables:

$$\begin{aligned} N_k&=c_k^*c_k\,,\quad N_{-k}=c_{-k}^*c_{-k}\,,\\ \sigma_k&\equiv c_k c_{-k}\,,\quad \sigma_k\equiv c_k^*c_{-k}^*\,. \end{aligned} \tag{4.29}$$

As a result one obtains

$$\left[i\left(\frac{d}{dt}+\Gamma\right)+(\omega_p-\omega_R)\right]\tilde{R}=F-\frac{iG_k}{2}\sigma_k\,, \tag{4.30}$$

$$\left[i\left(\frac{d}{dt}+2\eta_k\right)+(\omega_\text{p}-2\omega_k)\right]\sigma_k = i\frac{G_k}{2}\tilde{R}(N_k+N_{-k})\,, \tag{4.31}$$

$$\left(\frac{d}{dt}+2\eta_k\right)N_k = \frac{G_k}{2}\left(\tilde{R}\sigma_k^* + \tilde{R}^*\sigma_k\right)\,, \tag{4.32}$$

$$\left(\frac{d}{dt}+2\eta_k\right)(N_k-N_{-k}) = 0\,. \tag{4.33}$$

Here we have used the inversion symmetry: $G_k = G_{-k}$, $\omega_k = \omega_{-k}$ and $\eta_k = \eta_{-k}$. From (4.33) it then follows that

$$N_k = N_{-k} \quad \text{at} \quad t \gg 1/2\eta_k\,. \tag{4.34}$$

Combining (4.31) (and its complex conjugate) and (4.32), one can obtain the relation

$$\left(\frac{d}{dt}+4\eta_k\right)\left(N_k^2-|\sigma_k|^2\right) = 0\,. \tag{4.35}$$

Thus

$$|\sigma_k| = N_k \quad \text{at} \quad t \gg \frac{1}{4\eta_k} \tag{4.36}$$

for any solution of the above system of (4.30)–(4.33). It is convenient to introduce a phase angle θ_k as

$$\sigma_k = N_k \exp\left[i\left(\theta_k - \frac{\pi}{2}\right)\right] = -iN_k e^{i\theta_k} \tag{4.37}$$

and rewrite the dynamic equations in the form

$$\left[i\left(\frac{d}{dt}+\Gamma\right)+(\omega_\text{p}-\omega_\text{R})\right]\tilde{R} = F - \frac{G_k}{2}N_k e^{i\theta_k}\,, \tag{4.38}$$

$$\frac{d}{dt}\theta_k - (\omega_\text{p}-2\omega_k) - \frac{G_k}{2}\left(\tilde{R}e^{-i\theta_k}+\tilde{R}^*e^{i\theta_k}\right) = 0\,, \tag{4.39}$$

$$\left[\frac{d}{dt}+2\eta_k - i\frac{G_k}{2}\left(\tilde{R}e^{-i\theta_k}-\tilde{R}^*e^{i\theta_k}\right)\right]N_k = 0\,. \tag{4.40}$$

In the stationary ($d/dt = 0$, $\omega_\text{p} = 2\omega_k$) state from (4.38)–(4.40) we obtain

$$\tilde{R}^{(0)} = \frac{F-(G_k/2)N_k^{(0)}\exp(i\theta_k^{(0)})}{i\Gamma+(\omega_\text{p}-\omega_\text{R})} \tag{4.41}$$

and

$$\tilde{R}^{(0)}\exp\left(-i\theta_k^{(0)}\right) + \tilde{R}^{*(0)}\exp\left(i\theta_k^{(0)}\right) = 0 \tag{4.42}$$

$$\tilde{R}^{(0)}\exp\left(-i\theta_k^{(0)}\right) - \tilde{R}^{*(0)}\exp\left(i\theta_k^{(0)}\right) = -i\frac{4\eta_k}{G_k}\,. \tag{4.43}$$

Substituting relation (4.41) and its complex conjugate into (4.42) and (4.43), one can represent the equations of the stationary state in the form, which resembles stationary equations of S theory

$$h\tilde{V}_k \sin\left(\theta_k^{(0)} - \theta_{\mathrm{R}}\right) = -\mathcal{S}_{\mathrm{R}} N_k^{(0)} , \tag{4.44}$$

$$h\tilde{V}_k \cos\left(\theta_k^{(0)} - \theta_{\mathrm{R}}\right) = \eta_k + \eta_{\mathrm{rd}} N_k^{(0)} . \tag{4.45}$$

Here

$$h\tilde{V}_k \equiv \frac{F G_k}{2\sqrt{\Gamma^2 + \Delta\omega_{\mathrm{R}}^2}} \tag{4.46}$$

describes an effective pumping, $\Delta\omega_{\mathrm{R}} = \omega_{\mathrm{p}} - \omega_{\mathrm{R}}$,

$$\mathcal{S}_{\mathrm{R}} \equiv \frac{\Delta\omega_{\mathrm{R}} G_k^2}{4\left(\Gamma^2 + \Delta\tilde{\omega}_{\mathrm{R}}^2\right)} , \tag{4.47}$$

$$\eta_{\mathrm{rd}} = \frac{\Gamma G_k^2}{4\left(\Gamma^2 + \Delta\omega_{\mathrm{R}}^2\right)} , \tag{4.48}$$

describe effective nonlinearities responsible for the phase mismatching ($\mathcal{S}_{\mathrm{R}}$) and nonlinear radiation damping (η_{rd}), respectively:

$$\sin\theta_{\mathrm{R}} \equiv \frac{\Delta\omega_{\mathrm{R}}}{\sqrt{\Gamma^2 + \Delta\omega_{\mathrm{R}}^2}} ,$$

$$\cos\theta_{\mathrm{R}} \equiv \frac{\Gamma}{\sqrt{\Gamma^2 + \Delta\omega_{\mathrm{R}}^2}} . \tag{4.49}$$

The threshold of parametric pumping, as it follows from (4.45), is defined by

$$h_{\mathrm{c}}\tilde{V}_k = \eta_k . \tag{4.50}$$

The total number of excited magnon pairs above the threshold is defined by the following relation:

$$(h\tilde{V}_k)^2 = \left(\mathcal{S}_{\mathrm{R}} N_k^{(0)}\right)^2 + \left(\eta_k + \eta_{\mathrm{rd}} N_k^{(0)}\right)^2 . \tag{4.51}$$

Thus, we see that the restriction of the magnon pair growth arises from the nonlinear coupling with the microwave resonator. In the simplest model the microwave magnetic field in a resonator cavity of the volume V_{R} is given by

$$h(t) \simeq i\sqrt{\frac{2\pi\hbar\omega_{\mathrm{p}}}{V_{\mathrm{R}}}}(R - R^*) .$$

Comparing terms $i(G_k/2)Rb_k^* b_{-k}^*$ and $h(t)V_k b_k^* b_{-k}^*$, which describe excitation of parametric magnon pairs, we obtain

$$\frac{G_k}{2} \simeq \sqrt{\frac{2\pi\hbar\omega_{\mathrm{p}}}{V_{\mathrm{R}}}} V_k \tag{4.52}$$

and finally

$$\eta_{\rm rd} \simeq \frac{4\pi\hbar Q V_k^2}{V_{\rm R}} , \tag{4.53}$$

where $Q = \omega_{\rm R}/2\Gamma$ is the quality factor of the resonator.

4.3 Nonlinear SR Theory

As we have seen above, the principal element of the parametric resonance theory is the magnon pair (for example [77], see also Appendix D), which appears as a result of nonlinear excitation of the sample by the pump field. Mathematically the magnon pair consists of the following terms:

$$\begin{aligned} K_-(\boldsymbol{k}) &= b_k b_{-k} , \\ K_+(\boldsymbol{k}) &= b_k^* b_{-k}^* , \\ K_0(\boldsymbol{k}) &= \frac{1}{2}(b_k^* b_k + b_{-k}^* b_{-k}) . \end{aligned} \tag{4.54}$$

Using the classical analog of commutator, we obtain

$$\begin{aligned} [K_0(\boldsymbol{k}), K_\pm(\boldsymbol{q})]_c &= \pm K_\pm(\boldsymbol{k})\Delta(\boldsymbol{k}-\boldsymbol{q}) , \\ [K_-(\boldsymbol{k}), K_+(\boldsymbol{q})]_c &= 2K_0(\boldsymbol{k})\Delta(\boldsymbol{k}-\boldsymbol{q}) . \end{aligned} \tag{4.55}$$

Strictly mathematically speaking, we can consider all three terms in (4.54) for a given $\boldsymbol{k}$ as the classical generators of $SU(1,1)$ algebra with the Casimir invariant (see, for example, [11])

$$C(\boldsymbol{k}) = K_0^2(\boldsymbol{k}) - \frac{1}{2}\left[K_+(\boldsymbol{k})K_-(\boldsymbol{k}) + K_-(\boldsymbol{k})K_+(\boldsymbol{k})\right] . \tag{4.56}$$

The effective Hamiltonian of the combined (resonator and a parametric pair of magnons) system can be written as follows

$$\begin{aligned} \frac{\mathcal{H}}{\hbar} &= \omega_{\rm R} R^* R + 2\omega_k K_0(\boldsymbol{k}) \\ &\quad + \frac{iG_k}{2}\left[R K_+(\boldsymbol{k}) - R^* K_-(\boldsymbol{k})\right] + \frac{\mathcal{H}_{\rm int}}{\hbar} + \frac{\mathcal{H}_t}{\hbar} \end{aligned} \tag{4.57}$$

where

$$\begin{aligned} \frac{\mathcal{H}_{\rm int}}{\hbar} &= 2\mathcal{T}_k K_0(\boldsymbol{k}) K_0(\boldsymbol{k}) + \mathcal{S}_k K_+(\boldsymbol{k}) K_-(\boldsymbol{k}) \\ &\quad + 2\mathcal{T}_k^{(\rm R)} R^* R K_0(\boldsymbol{k}) + \Phi_{\rm R} R^* R^* R R \end{aligned} \tag{4.58}$$

is the diagonal in the magnon pairs Hamiltonian of interaction; $\mathcal{T}_k$ and $\mathcal{S}_k$ are the amplitudes of magnon scattering; $\mathcal{T}_k^{(\rm R)}$ describes the effective photon–magnon interaction and $\Phi_{\rm R}$ the photon–photon scattering. The effective interactions in (4.58) arise as a result of the exclusion of nonresonance terms from the Hamiltonian of a magnetic system located in the resonator by nonlinear canonical transformations (see Appendix C).

$$\frac{\mathcal{H}_t}{\hbar} = \left[F \cdot R^* \exp(-i\omega_{\rm p} t) + F_1 \cdot R^* \exp(-i\omega t) + \text{c.c.}\right] + 2\frac{\partial \omega_k}{\partial H} H_{\rm m} \cos(\omega_{\rm m} t) K_0(\boldsymbol{k}) , \tag{4.59}$$

where $F\ (= F^*)$ is the parameter of the pumping field; and F_1 the parameter of the probe microwave field of the frequency ω. The last term in (4.59) describes the radio-frequency modulation of the magnon spectrum. The dynamic equations of motion with damping parameters Γ for R and $2\eta_k$ for $K_\nu(\boldsymbol{k})$ can be written in the form

$$i\left(\frac{d}{dt} + \Gamma\right) R = \frac{\partial(\mathcal{H}/\hbar)}{\partial R^*} ,$$
$$i\hbar\left(\frac{d}{dt} + 2\eta_k\right) K_\nu(\boldsymbol{k}) = [K_\nu(\boldsymbol{k}), \mathcal{H}]_c , \tag{4.60}$$

where $\nu = \pm, 0$. As a result one obtains

$$i\left(\frac{d}{dt} + \Gamma\right) R = \tilde{\omega}_{\rm R} R + F e^{-i\omega_{\rm p} t} + F_1 e^{-i\omega t} - \frac{i G_k}{2} K_-(\boldsymbol{k}) , \tag{4.61}$$

$$i\left(\frac{d}{dt} + 2\eta_k\right) K_-(\boldsymbol{k}) = 2\left[\tilde{\omega}_k + \frac{\partial \omega_k}{\partial H} H_{\rm m} \cos(\omega_{\rm m} t)\right] K_-(\boldsymbol{k}) + \left[i G_k R + 2\mathcal{S}_k K_-(\boldsymbol{k})\right] K_0(\boldsymbol{k}) , \tag{4.62}$$

$$\left(\frac{d}{dt} + 2\eta_k\right) K_0(\boldsymbol{k}) = \frac{G_k}{2}\left[R K_+(\boldsymbol{k}) + R^* K_-(\boldsymbol{k})\right] . \tag{4.63}$$

Here

$$\tilde{\omega}_{\rm R} = \omega_{\rm R} + 2\mathcal{T}_k^{(\rm R)} K_0(\boldsymbol{k}) + 2\Phi_{\rm R} |R|^2 ,$$
$$\tilde{\omega}_k = \omega_k + 2\mathcal{T}_k K_0(\boldsymbol{k}) + \mathcal{T}_k^{(\rm R)} |\tilde{R}|^2 , \tag{4.64}$$

are the effective frequencies of the resonator mode and quasi phonon, respectively, which describe nonlinear frequency shifts in the excited system.

We can introduce slow variables

$$K_-(\boldsymbol{k}) = \sigma_k \exp(-i\omega_{\rm p} t) ,$$
$$R = \tilde{R} \exp(-i\omega_{\rm p} t) ,$$

and obtain the following equations of motion:

$$\left[i\left(\frac{d}{dt} + \Gamma\right) + (\omega_{\rm p} - \tilde{\omega}_{\rm R})\right] \tilde{R} = -\frac{i G_k}{2} \sigma_k + F + F_1 e^{i(\omega_{\rm p} - \omega)t} , \tag{4.65}$$

$$\left\{ i\left(\frac{d}{dt}+2\eta_k\right)+\left[\omega_p-2\tilde{\omega}_k-2\frac{\partial\omega_k}{\partial H}H_m\cos(\omega_m t)\right]\right\}\sigma_k = N_k\mathcal{P}_k\,, \tag{4.66}$$

$$\left(\frac{d}{dt}+2\eta_k\right)N_k=\frac{G_k}{2}\left(\tilde{R}\sigma_k^*+\tilde{R}^*\sigma_k\right)\,. \tag{4.67}$$

Here

$$\mathcal{P}_k = iG_k\tilde{R}+2\mathcal{S}_k\sigma_k$$

is the effective photon–quasi phonon coupling parameter, and $N_k \equiv K_0(\boldsymbol{k})$.

Combining (4.66) (with its complex conjugate) and (4.67), we obtain

$$\left(\frac{d}{dt}+4\eta_k\right)\left(N_k^2-|\sigma_k|^2\right)=0\,. \tag{4.68}$$

Thus $|\sigma_k| = N_k$ for any solution of (4.65)–(4.67) at $t \gg 1/4\eta_k$. The Casimir invariant in this case is $C(\boldsymbol{k}) = 0$.

As above, we introduce the phase angle θ_k as

$$\sigma_k = N_k\exp\left[i\left(\theta_k-\frac{\pi}{2}\right)\right]\,. \tag{4.69}$$

The phase θ_k describes a deviation of forced oscillations of magnon pair from the pump field. We thus can rewrite the (4.65)–(4.67) in the form

$$\left[i\left(\frac{d}{dt}+\Gamma\right)+(\omega_p-\tilde{\omega}_R)\right]\tilde{R}=-\frac{G_k}{2}N_k e^{i\theta_k} + F + F_1 e^{i(\omega_p-\omega)t}\,, \tag{4.70}$$

$$\frac{d}{dt}\theta_k-\left(\omega_p-2\tilde{\omega}_k-2\frac{\partial\omega_k}{\partial H}H_m\cos(\omega_m t)\right) = \frac{G_k}{2}\left(\tilde{R}e^{-i\theta_k}+\tilde{R}^*e^{i\theta_k}\right)-2\mathcal{S}_k N_k \tag{4.71}$$

$$\left[\frac{d}{dt}+2\eta_k-i\frac{G_k}{2}\left(\tilde{R}e^{-i\theta_k}-\tilde{R}^*e^{i\theta_k}\right)\right]N_k=0\,. \tag{4.72}$$

The power P absorbed by the combined "resonator-sample" system from the microwave pumping field is defined by the following expression:

$$P=\frac{\partial\mathcal{H}_t}{\partial t}=2\hbar\omega_p F\cdot\mathrm{Im}\tilde{R}\,. \tag{4.73}$$

We shall calculate responses of this system to the probe microwave field and modulating RF field. These responses are defined by linearizing (4.70)–(4.72) on small deviations

$$\begin{aligned}\tilde{R}(t)&=\tilde{R}^{(0)}+\delta\tilde{R}(t)\,,\\ \tilde{R}^*(t)&=\tilde{R}^{*(0)}+\delta\tilde{R}^*(t)\,,\\ \theta_k(t)&=\theta_k^{(0)}+\delta\theta_k(t)\,,\\ N_k(t)&=N_k^{(0)}+\delta N_k(t)\,,\end{aligned} \tag{4.74}$$

in the vicinity of the stationary state defined by

$$\tilde{R}^{(0)} = \frac{F - (G_k/2) N_k^{(0)} \exp(i\theta_k^{(0)})}{i\Gamma + (\omega_p - \tilde{\omega}_R)} \tag{4.75}$$

and

$$\tilde{R}^{(0)} \exp(-i\theta_k^{(0)}) + \tilde{R}^{*(0)} \exp(i\theta_k^{(0)}) = \frac{4\mathcal{S}_k N_k^{(0)}}{G_k} , \tag{4.76}$$

$$\tilde{R}^{(0)} \exp(-i\theta_k^{(0)}) - \tilde{R}^{*(0)} \exp(i\theta_k^{(0)}) = -i\frac{4\eta_k}{G_k} . \tag{4.77}$$

As above, we can represent the equations of the stationary state in the form, which resembles stationary equations of S theory:

$$h\tilde{V}_k \sin\left(\theta_k^{(0)} - \theta_R\right) = -\left(\mathcal{S}_k + \mathcal{S}_R\right) N_k^{(0)} , \tag{4.78}$$

$$h\tilde{V}_k \cos\left(\theta_k^{(0)} - \theta_R\right) = \eta_k + \eta_{rd} N_k^{(0)} . \tag{4.79}$$

Here

$$h\tilde{V}_k \equiv \frac{F G_k}{2\sqrt{\Gamma^2 + \Delta\omega_R^2}} \tag{4.80}$$

is the effective pump field. The following coefficients

$$\mathcal{S}_R \equiv \frac{\Delta\omega_R G_k^2}{4(\Gamma^2 + \Delta\tilde{\omega}_R^2)} , \tag{4.81}$$

$$\eta_{rd} = \frac{\Gamma G_k^2}{4\left(\Gamma^2 + \Delta\omega_R^2\right)} , \tag{4.82}$$

are responsible for effective nonlinearities for the phase mismatching ($\mathcal{S}_R$) and nonlinear radiation damping (η_{rd}), respectively:

$$\sin\theta_R \equiv \frac{\Delta\omega_R}{\sqrt{\Gamma^2 + \Delta\omega_R^2}} ,$$

$$\cos\theta_R \equiv \frac{\Gamma}{\sqrt{\Gamma^2 + \Delta\omega_R^2}} . \tag{4.83}$$

The threshold of parametric pumping, is defined by

$$h_c\tilde{V}_k = \eta_k . \tag{4.84}$$

The total number of excited magnon pairs above the pumping threshold is

$$(h\tilde{V}_k)^2 = \left[\left(\mathcal{S}_k + \mathcal{S}_R\right) N_k^{(0)}\right]^2 + \left(\eta_k + \eta_{rd} N_k^{(0)}\right)^2 . \tag{4.85}$$

4.4 Experimental Techniques

As we have already discussed above, the language of harmonic oscillators makes it possible to consider quite different physical systems as systems with similar behavior. Studying the behavior of quasi particles of one sort, one can expect in general a similar behavior of quasi particles of the other sort (except very specific cases) if the Hamiltonians of both systems are similar to each other. We shall use this similarity applying the theory developed above to the case of quasi phonons in an antiferromagnet. Experimenting with parametric excitation of magnetoelastic waves at room temperature and in the range of microwaves below 4 GHz is relatively simple [79–83]. On the other hand, a general information we obtain on parametrically excited quasi phonon pairs is very valuable to figure out what kind of physics can be expected for excited magnon pairs that usually have higher frequencies. Thus, a more complicated microwave technique should be used in this case.

Parametric excitation of quasi phonons was investigated using a decimeter-band spectrometer (see, for example, [34]). Open helical resonator, which is a half-wavelength dipole coiled into a helix at the ends of which reflection of an electromagnetic wave occurs, was used as a measuring unit (see, Figure 4.2).

The inner diameter of the helix equals 0.5 cm and the diameter of the copper wire was 0.5 mm. To a first approximation (without allowance for capacitances between winding coils), the wire length needed to make the helix is $l = \lambda/2$. In such a resonance structure the microwave magnetic field $\boldsymbol{h}(t)$ was directed along the axis of the helix. The quality factor of this resonator at liquid nitrogen temperature was about $Q \sim 300$.

The capacitance coupling of the coaxials with the helix is achieved by input antennas 1 and 2 and output antenna 3. In contrast to many other experiments on parametric excitation of waves in magnetic materials in which the input coupling was set close to critical value ($\beta \sim 1$), we used weak coupling for all external microwave circuits in order to decrease their effect on the cavity. The microwave power from the generator for parametric excitation of quasi phonons in the sample was fed through the first antenna. The sweep signal from the probe microwave generator, which did not excite parametric resonance (the response to this signal was always linear), was fed from the second antenna. This signal was used to investigate the spectrum of the coupled photon–quasi phonon oscillations in the combined res-

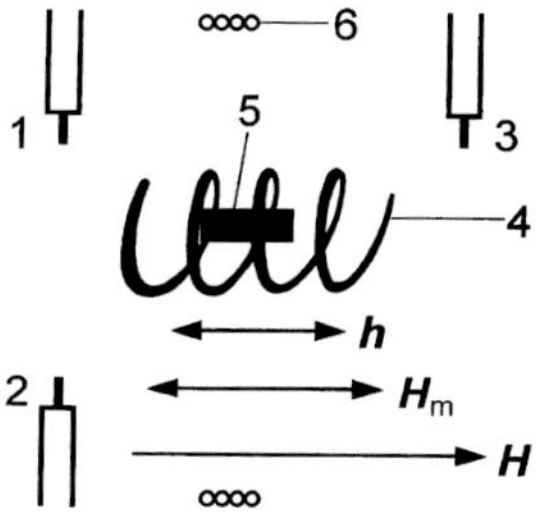

Figure 4.2 Schematic diagram of the absorbing cell and the direction of the magnetic fields: coaxial waveguides from the pump generator (1), the probe generator (2), and the receiving channel (3), respectively, and helical resonator (4), sample (5), modulation coil (6).

onator-sample system. The signal from the received antenna was fed into a spectrum analyzer and the receiver. Measured in such a way, the spectrum showed the amplitude-frequency characteristics of the studied resonator–sample system.

The emission of electromagnetic waves by parametrically excited quasi phonons was also investigated. The emission was observed after the end of the microwave pump pulse. In these cases the microwave signal passed through the cavity was detected by the crystal detector and displayed on an oscilloscope.

Besides microwave fields, the RF magnetic field $H_{\mathrm{m}} \cos(\omega_{\mathrm{m}} t)$ with the frequency $\omega_{\mathrm{m}} = 2\pi \cdot (0.1 - 2)$ MHz was applied to the sample for weak modulation of the static magnetic field, and hence the spectrum of magnetoelastic waves. The modulation field at the sample was created by a coil of 2 cm diameter situated coaxially with the helix resonator. The static ($\boldsymbol{H}$), microwave ($\boldsymbol{h}$) and modulation ($\boldsymbol{H}_{\mathrm{m}}$) magnetic fields were parallel to each other and were presented in the basal plane of the crystal being investigated.

When the spectrum of parametrically excited quasi phonons is modulated by an oscillating magnetic field $H_{\mathrm{m}} \cos(\omega_{\mathrm{m}} t)$ the amplitude of microwave pump signal which passes through the resonator with the sample is modulated at the frequency ω_{m} [87]. This frequency mixing appears just above the threshold of pumping. The so-called modulation response of the sample was studied in present papers by two methods: (i) by measuring the amplitudes of the side bands at the frequencies $\omega_{\mathrm{p}} \pm \omega_{\mathrm{m}}$ on the screen of the spectrum analyzer, and (ii) by measuring the amplitude of modulation of the pump power passed through the helix resonator on the output of microwave receiver.

All measurements were performed on an antiferromagnetic single-crystal sample of $FeBO_3$. This crystal has rhombohedral symmetry D_{3d}. After antiferromagnetic ordering the magnetic moments of the sublattices lie in the basal plane perpendicular to the threefold "hard" axis. The Dzyaloshinskii–Moriya interaction leads to a " sloping" of the spins, as a result of which there appears a weak ferromagnetic moment lying in the basal plane. The samples used were naturally faceted laminas of thickness $\sim$ 0.1 cm and linear dimensions 0.3–0.7 cm. The basal plane coincided with the planes of the laminas. Each sample was fitted into a teflon holder using a bag made of teflon film, and the whole assembly was placed inside the resonator. The helix with the sample was placed in gaseous nitrogen. Using single crystals of volume 0.01–0.02 cm^3, one obtained a filling factor for the sample in the helix resonator up to 5. This increased the efficiency of the resonator-sample coupling and made it possible to observe coupled photon–quasi phonon modes at a low supercriticality.

Experiments were carried out at the pumping frequency $\omega_{\mathrm{p}} \sim 2\pi \cdot 1$ GHz, at magnetic field $H = 30$–500 Oe (SI: 2.4–40 kA/m), and at liquid nitrogen (77 K) and room (293 K) temperatures. Both pulsed and continuous microwave pumping were used to excite the parametric magnetoelastic waves in the sample. The pulses had 100–1500 μs duration and the repetition frequency of 50 Hz. Only pairs of magnetoelastic waves (quasi phonons) were excited in our experiments as far as the bottom of the spin-wave spectrum was always much greater than the pumping frequency. The threshold amplitude of the parallel pumping of quasi phonons was recorded according to the characteristic distortion in the shape of the pulse passed through the resonator.

4.5 Experimental Results

Typical linear responses of the combined system of resonator and sample on a weak signal of probe generator are shown in Figures 4.3 and 4.4. Figures 4.3(a) and 4.4(a) show the line shape of resonator and sample system for two resonator frequencies in the case when the input microwave power from the pump generator is equal to zero ($P_{in} = 0$). These data are in agreement with the Lorentzian-shaped lines (solid lines Figures 4.3(a) and 4.4(a)). Such line shape of the system remains the same till the pump power reaches the threshold of parametric instability with the critical value P_c.

Just above the critical value of the input power ($P > P_c$), the line shape of the system begins to change and deviate from the Lorentzian form. There is a characteristic power P_{sp} at which the line shape splits. In this experiment it was not possible to determine accurately the point of splitting and details of the line shape in the frequency range $|\omega - \omega_p| < 2\pi \cdot 0.2\,\text{MHz}$ due to the presence of a powerful microwave signal from the pump generator. However it was observed that P_{sp} increases with decreasing volume of the sample.

The typical experimental line shape of the resonator and sample system is shown in Figure 4.3(b, c) and Figure 4.4(c). In these cases the pump frequency ω_p is equal to the frequency ω_R of the helix resonator. It was observed that the splitting distance between two peaks increases with increasing pump power. This fact exhibits a characteristic repulsion of the spectra of the normal modes in the system of coupled oscillators. Analogous results were obtained for the whole range of static magnetic field and for both temperatures (77 K and 293 K).

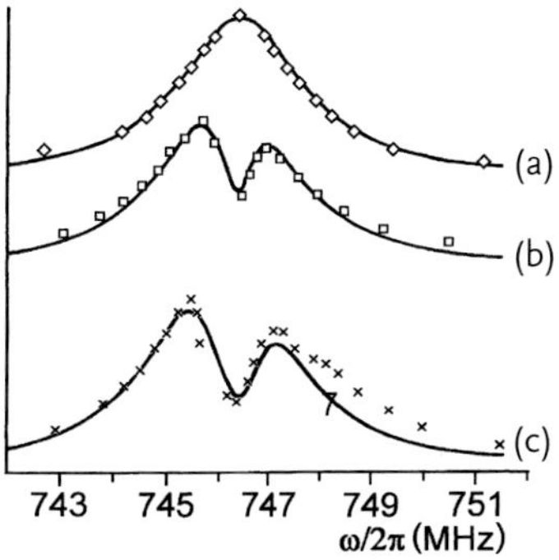

Figure 4.3 Spectrum of the probe signal passed through the resonator: (a) below the threshold of parametric excitation, (b) and (c) above the threshold. (b) $P_{in}/P_c = 5.1$; (c) $P_{in}/P_c = 29.7$, $H = 144$ Oe (SI: 11.5 kA/m), $T = 77$ K, $\omega_R = \omega_p = 2\pi \cdot 746$ MHz. The solid lines demonstrate the fitting of data using the equivalent circuit.

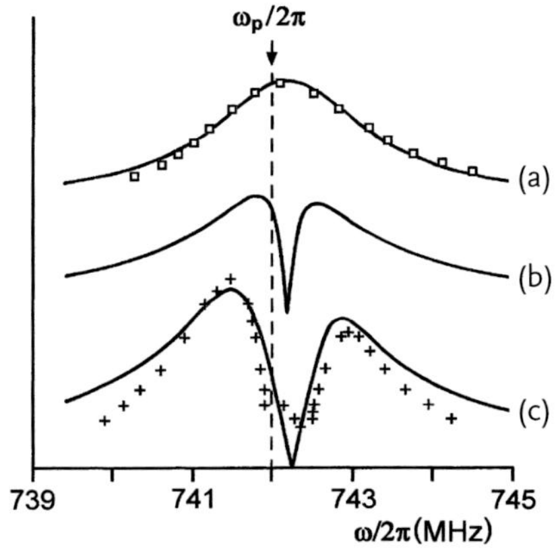

Figure 4.4 Spectrum of the probe signal passed through the resonator: (a) the pump power $P_{in} = 0$ (the solid line is a Lorentzian), and (c) $P_{in}/P_c = 30.6$, $H = 144$ Oe (SI: 11.5 kA/m), $T = 77$ K, $Q_R = 380$, $\omega_R = 2\pi \cdot 742.2$ MHz, $\omega_p = 2\pi \cdot 742$ MHz. The solid lines (b) and (c) correspond to the theory (see the text in the theoretical section).

It is reasonable to associate this repulsion with the threshold onset of nonlinear coupling of two oscillators: electromagnetic field oscillation in the resonator cavity and the parametric pair of quasi phonons in the sample [77].

4.5.1
Equivalent Circuit

Traditionally the microwave resonator was considered as the LC circuit. The small sample in the cavity was taken into account as an additional active resistance R which depends on the absorbed power [84]. Increase of R leads to the diminishing of quality factor of the resonator and shifts the resonance frequency. The value of R and the absorbed microwave power usually were calculated for a given pump power. The line shape of the resonator and sample system was considered to be Lorentzian.

It is obvious that this simple scheme does not take into account the inverse effect of the excited sample on the resonator. In order to obtain a two-hump line shape it is necessary to replace simple active resistance R by an additional LC circuit. Consider the equivalent circuit (see, Figure 4.5) which corresponds to the case of two coupled oscillators with frequencies $\omega_R = (LC)^{-1/2}$ and $\omega_s = (L_s C_s)^{-1/2}$, $\omega_R \simeq \omega_s$. The input and output are connected by long lines with the transformation coefficients M_1 and M_2. We can write the Kirchhoff equations with the input

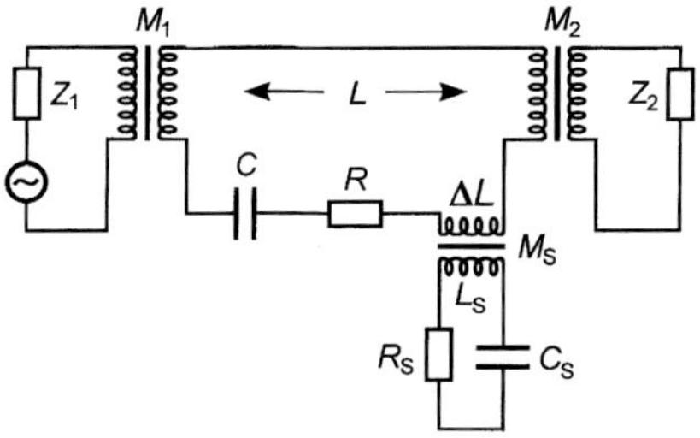

Figure 4.5 Equivalent electric circuit of the resonator with the sample.

harmonic signal $E\exp(i\omega t)$:

$$\begin{aligned}
E - i\omega M_1 I_c &= I_1 z_1 \,, \\
-i\omega M_2 I_c &= I_2 z_2 \,, \\
-i\omega M_1 I_1 - i\omega M_2 I_2 - i\omega M_s I_s &= I_c\left(R + i\omega L - \frac{i}{\omega C} + i\omega \Delta L\right) \,, \\
-i\omega M_s I_c &= I_s\left(R_s + i\omega L_s - \frac{i}{\omega C_s}\right) \,.
\end{aligned} \tag{4.86}$$

From these equations follows the ratio of the output voltage U_2 to the amplitude of the input signal E:

$$\frac{U_2}{E} = -\frac{\zeta\xi^3}{D}\sqrt{\frac{|z_1/z_2|}{Q_1 Q_2}}\exp(-i\phi_1) \,, \tag{4.87}$$

where

$$\begin{aligned}
D &= \frac{\xi}{\zeta Q_\mathrm{R}} + \zeta\xi^3\left[\frac{\exp(-i\phi_1)}{Q_1} + \frac{\exp(-i\phi_2)}{Q_2}\right] \\
&\quad + \frac{\zeta\xi^5}{Q_s^2 Q_\mathrm{Rs} D_\zeta} + i\left(\xi\frac{\omega_\mathrm{R}}{\tilde{\omega}_\mathrm{R}}\right)^2 - \frac{i}{\zeta^2} - i\frac{\zeta\xi^4(\xi^2-1)}{Q_s Q_\mathrm{Rs} D_\zeta} \,, \\
D_\zeta &= \left(\frac{\xi}{Q_s}\right)^2 + (\xi^2-1)^2 \,, \quad \xi \equiv \frac{\omega}{\omega_s} \,, \quad \zeta \equiv \frac{\omega_s}{\omega_\mathrm{R}} \,.
\end{aligned}$$

Here $Q_j \equiv L|z_j|/\omega_\mathrm{R} M_j^2$ $(j = 1, 2)$ are the quality factors of input and output circuits and $\exp(i\phi_j) \equiv z_j/|z_j|$; Q_R, Q_s are the quality factors of the resonator, sample, respectively; $Q_\mathrm{Rs} \equiv LR_s/\omega_\mathrm{R} M_\mathrm{Rs}^2$, $\tilde{\omega}_\mathrm{R} \equiv [(L + \Delta L)C]^{-1/2}$.

The solid lines in Figure 4.3 demonstrate the frequency dependence of

$$\frac{P_{out}}{P_{in}} = \left(\frac{U_2^2}{E^2}\right)\left(\frac{R_1}{R_2}\right), \tag{4.88}$$

at $z_j = R_j$ $(j = 1, 2)$. Figure 4.3(a) shows the case when the resonator and sample coupling is absent ($Q_{Rs}^{-1} = 0$ and $\omega_R/\tilde{\omega}_R = 1$). The following parameters were used: $Q_R = 290$ and $Q_1^{-1} + Q_2^{-1} \simeq 5 \times 10^{-4}$. Figure 4.3(b, c) describe the line shape of the system at $(\omega_R/\tilde{\omega}_R)^2 - 1 \leq 10^{-3}$ and $Q_s = 1300 \pm 200$, $Q_{Rs} = 600 \pm 60$, and $Q_s = 850 \pm 150$, $Q_{Rs} = 420 \pm 40$, respectively . The quality factors of the resonator and sample are given from the experiment on parametric excitation of quasi phonons in $FeBO_3$. We can see a good agreement between theory and experiment.

We can also write a useful formula for calculating the power absorbed by the sample relative to the input microwave power:

$$\frac{P_s}{P_{in}} = \frac{\zeta^2 \xi^8}{Q_1 Q_{Rs} Q_s^2 D_\zeta |D|^2}. \tag{4.89}$$

As it follows from the above equations, the proposed equivalent circuit can be simplified to a conventional small-sample approximation. This corresponds to the case when the circuit is so weakly coupled with the resonator that its effect on the resonator as a small active resistance (the resonator line shape still has the one-hump form). It is obvious that in the parametric resonance experiments the small-sample approximation is restricted for any size of the sample as far as the coefficient of coupling Q_{Rs}^{-1} increases with increasing pump power.

4.5.2 SR Theory and Experiment

For calculations we shall use the experimentally estimated damping $\eta_k = 2\pi \cdot 0.22$ MHz of quasi phonon and for simplicity we neglected the effective photon–photon scattering ($\Phi_R = 0$).

The solid line in Figure 4.4(c) demonstrates a good agreement for the theoretically calculated absorption $\delta P(t) \propto \mathrm{Im}\delta \tilde{R}(t)$ of the probe microwave field $F_1 \cdot \exp(-i\omega t)$ with the experimental data. It should be noted that the obtained phenomenological parameters have a definite physical meaning. The first parameter $G_k^2/2\Gamma|T_k| = 4 \pm 1$ characterizes the ratio of two mechanisms of restriction of absorbed power (nonlinear radiation damping and phase mismatching). For example, for the case of parametrically excited magnons in antiferromagnetic $CsMnF_3$ and $MnCO_3$ [87] the corresponding estimated parameter was ~ 0.1–1. The second parameter $G_k F/\Gamma\eta_k = 2 \pm 0.3$, which characterizes overcriticality, is about the experimental value $(P_{in}/P_c)^{1/2} \simeq 3$. The theoretical calculations were found to be not so sensitive to the third parameter within $-1 \leq T_k^{(R)}/T_k \leq 7$, ($T_k^{(R)}/T_k = 5$ in Figure 4.4).

The solid curve in Figure 4.4(b) is the theoretical calculation with the same parameters as for Figure 4.4(c), but at lower pumping $G_k F/\Gamma\eta_k = 1.3$. Thus the

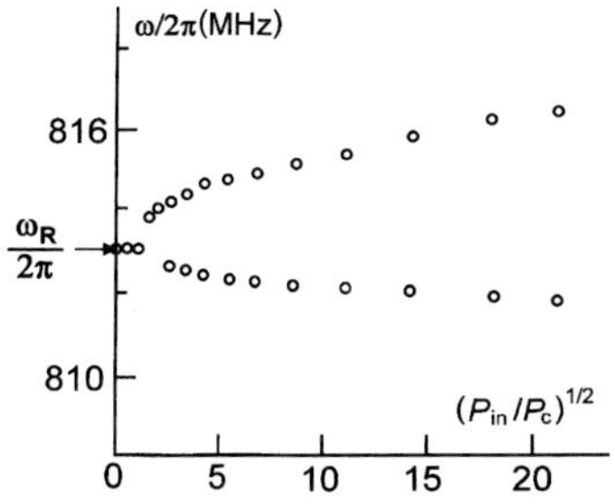

Figure 4.6 Frequency of normal modes in the system of coupled photon and pair of quasi phonons as a function of the amplitude of the microwave pump; $\omega_p = \omega_R = 2\pi \cdot 813$ MHz, $T = 77$ K, $H = 143$ Oe (SI: 11.4 kA/m).

curves Figure 4.4(b, c) demonstrate the dependence of splitting of photon–quasi phonon coupled modes on overcriticality $G_k F/\Gamma\eta_k > 1$.

The fact of repulsion of spectra can be naturally explained by an increase of coefficient of coupling between the electromagnetic mode of resonator and parametric pair of magnetoelastic waves with increasing P_{in}/P_c. Results for the frequencies of coupled modes in the resonator and sample system versus the microwave pump power are shown in Figure 4.6.

Figure 4.7 demonstrates two typical spectra for coupled (photon and pair of quasi phonons) modes in the cases, when there is a difference between the pump (ω_p) and the resonator (ω_R) frequencies. Measuring the maxima of these absorption lines at different ω_p, one can obtain the spectra of coupled oscillation versus pump frequency with the constant pump power (see, Figure 4.8).

Note that in the presented frequency range the pump power is always higher than the threshold power. It is obvious that if the frequency of the pump microwave field is far from the resonator frequency, then resonator–sample interaction is weak: the frequency of one mode is approximately equal to the pump frequency and the

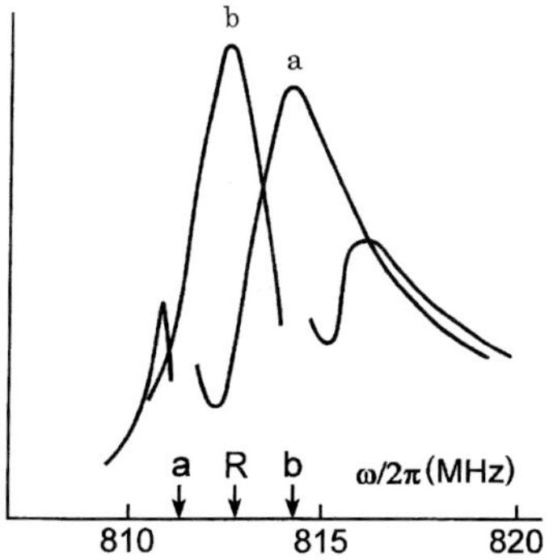

Figure 4.7 Spectrum of the probe signal passed through the resonator in the cases when the pump frequency is not equal to the resonator frequency. $\omega_R = 2\pi \cdot 813.4$ MHz, $T = 77$ K, $H = 144$ Oe (SI: 11.5 kA/m), $P_{in}/P_c = 125$. (arrow a) $\omega_p = \omega_a = 2\pi \cdot 811.9$ MHz, and (arrow b) $\omega_p = \omega_b = 2\pi \cdot 814.9$ MHz.

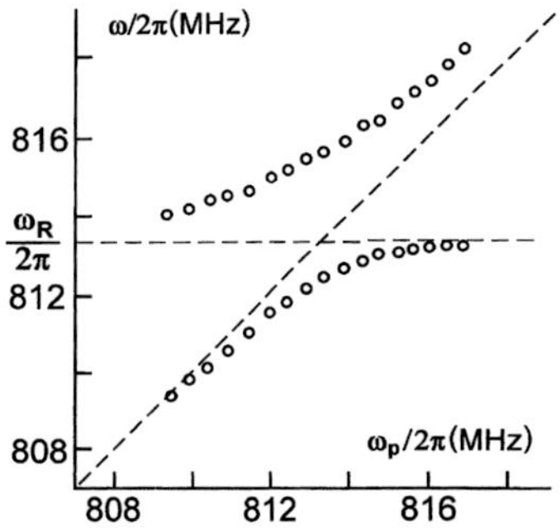

Figure 4.8 Frequency of normal modes versus the microwave pump frequency with constant pump power; $\omega_R = 2\pi \cdot 813.4$ MHz, $T = 77$ K, $H = 144$ Oe (SI: 11.5 kA/m), $P_{in}/P_c = 125$.

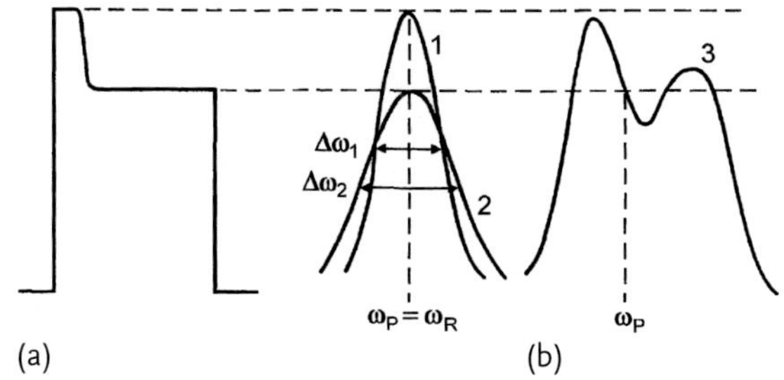

Figure 4.9 (a) Oscillograms of the microwave pulse after passage through the resonator. Distortion in the shape of the pulse corresponds to the beginning of the parametric excitation and the repulsion of the spectra of coupled oscillations. (b) Resonance absorption curve of the resonator with the sample below the beginning of parametric excitation (1); imaginary absorption curve of the resonator with the sample which corresponds to the conventional small-sample approximation (2); measured absorption curve shape of the resonator-sample system in the presence of the parametric excitation of quasi phonons (3).

frequency of the other mode is approximately equal to the resonator frequency. As the frequencies ω_p and ω_R converge, the increasingly stronger repulsion of the branches of the mixed photon–quasi phonon modes is observed. Thus, from the obtained results we can conclude, that the new normal modes (coupled photons and quasi phonon pairs) appear due to nonlinear interaction between the microwave resonator and the sample. The frequency of one mode is lower, and that of the other mode is higher than the pump frequency. As a result, this leads to a restriction of microwave power passed through the resonator (see, Figure 4.9).

Such a mechanism of restriction has not yet been considered for parallel pumping experiments.

According to the standard approximation of a small sample (see, for example, [84]) which is usually used for interpretation of parallel pumping experiments, the restriction of the passed microwave power has to be associated with the decrease of the quality factor of the resonator owing to microwave power absorption by the sample (see curves 1 and 2 in Figure 4.9b). As we can see, the main cause of the reduction of passed microwave power is quite different. It is due to the splitting of the resonance peak (see the curve 3 in Figure 4.9). It should be also noted that the correct line shape of the resonator and sample system cannot be obtained by a simple sweeping the frequency of the pump generator. This method was used in [88] for studying the parallel pumped spin waves in an antiferromagnet. It is obvious that the nonlinear response of the system to such swept signals should essentially differ from the linear response of a parametrically excited system to weak signal of the probe oscillator.

4.5.2.1 Modulation Response

In order to obtain more detailed information about coupled photon–quasi phonon interactions, we studied the spectra of collective oscillation of parametrically excited pairs of quasi phonons using the modulation method [85–87]. The response of resonator and sample system on the RF field was measured by two different methods under the same experimental conditions as given in Figure 4.4(c). Figure 4.10 shows the measured amplitudes of the side bands with the frequencies $\omega_p \pm \omega_m$ as a function of the modulation frequency ω_m.

Note that the obtained modulation response is not a symmetric one. The asymmetric character of the modulation response is associated with the fact that the sample is placed into the resonator cavity. For this reason we study some modulation response of the coupled photon–quasi phonon system instead of modulation response of parametric pairs of quasi phonons only. Since the spectrum of coupled photon–quasi phonon modes is an asymmetric one with respect to the pump frequency ω_p, the modulation response of this system does not need to be symmetric. Measured modulation amplitudes of the pump power passed through the resonator are shown in Figure 4.11.

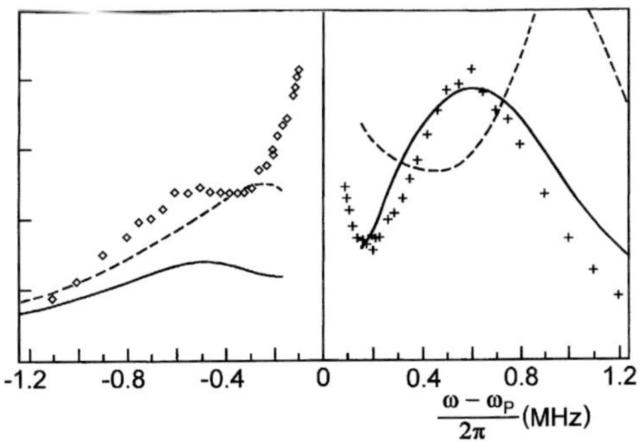

Figure 4.10 Measured amplitudes of the side bands with the frequencies $(\omega_p + \omega_m)$ and $(\omega_p - \omega_m)$ as a function of the modulation frequency ω_m. $\omega_p = \omega_R = 2\pi \cdot 742$ MHz, $P_{in}/P_c = 30.6$, $H = 144$ Oe (SI: 11.4 kA/m), $T = 77$ K, $Q_R = 310$. Abscissa of zero point corresponds to $\omega = \omega_p$. The maximum of observed response corresponds to the frequency $\omega_1 \sim 2\pi \cdot 600$ kHz. Theoretical curves (see Section 4.3) are shown by solid and dashed lines.

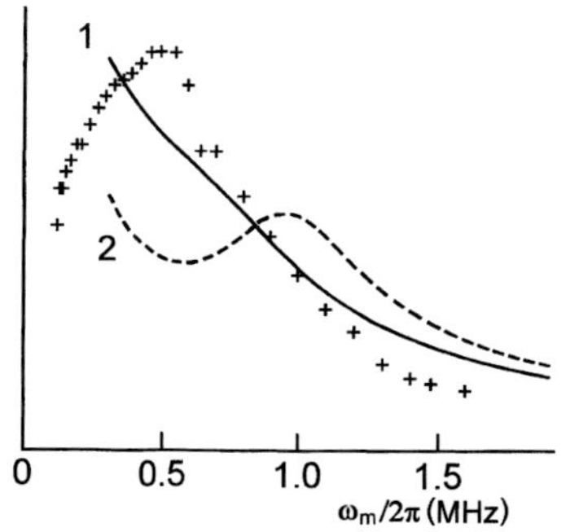

Figure 4.11 Measured modulation amplitude of the pump power passed through the resonator versus modulation frequency ω_m. Experimental conditions were the same as in Figure 4.10. The maximum value for modulation response corresponds to the frequency $\omega_2 \sim 2\pi \cdot 600$ kHz. Theoretical curves are shown by solid and dashed lines.

The solid line in Figure 4.10 demonstrates the response of a parametrically excited system to combined frequencies $\omega_p \pm \omega_m$. These signals appeared under RF modulation $H_m \cos \omega_m t$ (the experimental parameters are the same as in the case of Figure 4.4(c)). Qualitative agreement of the theory and experiment fails if we change the theoretical parameters, for example, if we take $G_k^2/2\Gamma|T_k| = 1$ (see, Figure 4.10b). Theoretically calculated modulation response

$$\delta P \propto \delta(N_k \cos \theta_k)$$

is shown by the solid curve in Figure 4.11. We have used the same parameters as those for Figure 4.4(c). If however we take $G_k^2/2\Gamma|T_k| = 1$ then we get the dashed curve in Figure 4.11, which differs much more from the experimental data than the solid line in Figure 4.11. We did not draw the theoretical curve in the range of small frequencies $\omega_m \leq 2\eta_k$, as far as we should take into account a slow motion of parametrically excited wave system in this case [87]. Such an effect has not been taken into account in this study.

The RF magnetic field modulates not only the spectrum of quasi phonons, but also the spectrum of coupled photon–quasi phonon modes (including their splitting). It is well known that the collective oscillations of parametric pairs represent the oscillation of their number and phase in the vicinity of equilibrium values. However, the efficiency of nonlinear photon–quasi phonon interaction depends on the number of parametric pairs. For this reason both the efficiency of photon–quasi phonon coupling and value of splitting oscillate with the same frequency and so the side band amplitudes in Figure 4.10 contain the information not only about parametrically excited magnetoelastic waves but also on the efficiency of nonlinear photon–quasi phonon interaction. A direct confirmation of existence of nonlinear coupled photon–quasi phonon modes were obtained by the observation of nonmonotonic electromagnetic emission from the resonator and sample system after the switching off of the pumping field [82]. The electromagnetic emission after the

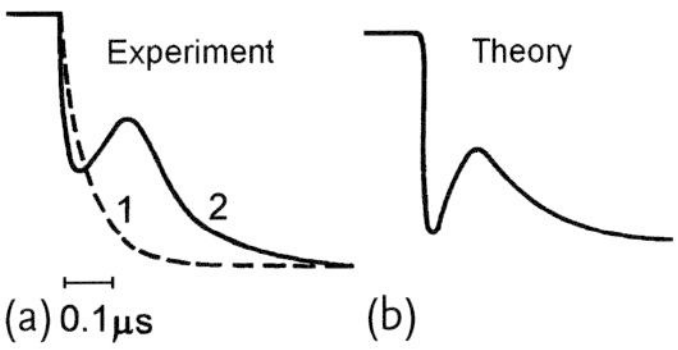

Figure 4.12 (a) Oscilloscope traces of the trailing edge of the microwave pump pulse passed through the resonator: for the case $h = 0.9h_c$ (1); $h \sim 2h_c$ (2, the ends of the two pulses have been superimposed). $H = 37$ Oe (SI: 2.94 kA/m), $T = 293$ K, $\omega_R = \omega_p = 2\pi \cdot 1217.4$ MHz. (b) Theoretically calculated (see Section 4.3) oscilloscope traces of the trailing edge of the microwave pulse passed through the resonator. $G_k F/\Gamma\eta_k = 2.2$; $T_k = S_k < 0$; $\eta_{rd}/|T_k| = 8$; $\eta_k = 2\pi \cdot 0.3$ MHz.

end of the microwave pump pulse gives the time evolution of this nonequilibrium system in the absence of the external driving force.

Figure 4.12 shows two representative oscilloscope traces of the trailing edge of the microwave pump pulse passed through the resonator. The microwave signal passed through the resonator was detected by a crystal detector and displayed on an oscilloscope. When the amplitude of the pump field is below the threshold, the signal from the detector on the trailing edge of the pulse decreases monotonically with characteristic relaxation time ~ 0.1 μs. The detected signal is proportional to the number of photons in the cavity, and the photon lifetime is determined by the quality factor of the resonator $Q \sim 300$. When the pump power exceeds the threshold P_c, we observe nonmonotonic emission from the resonator and sample after the trailing edge of the pulse.

First there is a sharp decrease in the microwave power transmitted through the cavity (the trailing edge of the pulse is steeper than below the threshold for excitation of the parallel pumping of quasi phonons). Then, the signal begins to rise. Its amplitude reaches a maximum at a time $\tau = 0.1$–0.5 μs after the end of the pulse. Then a monotonic decay of the emission intensity with a time scale ~ 1 μs is observed. The power output of emission has the same order of magnitude as the pump power passed through the resonator. Neither the time τ or other characteristics of the emission depend on the pulse length or repetition frequency of the microwave pulses. The only important point is that the pump pulse should be long enough for a parametric instability to occur in the sample and for a steady-state absorption level to be reached. We also measured the characteristic emission frequency ω_{em} and found that its value is slightly lower than the frequency of the pump field ω_p, $(\omega_p - \omega_{em}) \sim 1$ MHz. Taking into account the amplitude and the frequency of electromagnetic emission one may make the conclusion that photons are emitted by coherent parametric pairs of quasi phonons and not by individually excited quasi phonons.

Thus, we observed a kind of beating typical for the system of two coupled oscillators: (i) electromagnetic mode of the resonator, and (ii) parametric pair of magnetoelastic waves in the sample. The system of equations in (4.70)–(4.72) can describe the phenomenon of beating of coupled photon–quasi phonon modes. The result of numerical calculations of the behavior of the resonator and sample system after switching off the microwave pumping field are shown in Figure 4.12b. We can see both the qualitative and quantitative agreement between theory and experiment.

Thus we have obtained a good agreement between theory and experiment in the framework of the considered model. This is a basis for subsequent purposeful investigations of parametrically excited waves in a sample placed in a resonator. Coupled magnon–photon modes in YIG were studied in [89].

As it follows from our study, the nonlinear interaction between the resonator and the sample is very essential in the experiments of parallel microwave pumping of parametric quasi phonons. This interaction makes an interpretation of experimental results more complicated especially with respect to obtaining information about the nonlinearities of the magnetic system itself.

Thus we encountered two problems: (i) correct measurements of principal parameters of the studied system; (ii) properly taking into account the influence of the resonator on the properties of the system under consideration. These two problems were not considered in the framework of the conventional approach.

To solve the first problem it was necessary to overcome the framework of small-sample approximation, because the traditional approach leads to significant errors. Incorrect measurement of the microwave power absorbed by the sample also gives an incorrect calculation of nonlinear magnetic susceptibility, number of parametric pairs, coefficient of nonlinear dumping, coefficients of interactions, and so on. We have proposed above a variant of an equivalent circuit which can resolve this problem. However solving the first problem we do not yet obtain any information about the nonlinear magnetoelastic system, such as the intrinsic amplitudes of quasi phonon scattering. It is thus important and necessary to understand the role of nonlinearities of the resonator and quasi phonon system.

To solve the second problem it was necessary to develop a conventional model based on *S* theory (sample in the alternating magnetic field $h \cos \omega_\mathrm{p} t$) taking into account a specific role of the microwave resonator. As shown in Section 4.3, the new theory is in a good agreement with obtained experimental data and therefore can be considered as a good basis for new purposeful studies of parametric pairs in various nonlinear media.

There is a simple way to avoid all the above problems by excluding the resonator from microwave measurements. This however leads to an essential decreasing ($\sim 1/Q$) of both the sensitivity of measurements and the amplitude of the microwave field on the sample.

On the other hand, we can consider a new role for the "resonator-sample" system. This system makes it possible to control the properties of magnetic materials by the microwave cavity. For example, the phase and amplitude of the microwave signal can be easily varied during the time $\sim 1\ \mu$s.

It is interesting to note that the above-mentioned nonlinear effects can also take place in microwave lines which do not include a resonator. The role of resonator mode in this case can be any oscillation which has a nonlinear coupling with parametric waves. Wiese *et al.* [90] considered three-magnon nonlinearity including the magnetostatic mode of the sample and found that this nonlinearity causes additional nonlinear magnon damping similar to nonlinear radiation damping. Sharaevskii *et al.* [91] observed a nonlinear influence of a powerful microwave signal on the damping of a weak microwave signal with close frequency in the delay line of magnetostatic waves. The dependence of damping of weak signals on their frequency demonstrates peaks typical for the spectrum of coupled oscillations. In this case, a characteristic repulsion in the spectrum of nonlinear coupled oscillators occurs in the system of linearly excited magnetostatic waves in YIG film and parametric pair of spin waves, which was excited by the magnetostatic waves above a certain threshold.

Problem 4.1. Develop a nonlinear theory of parametric excitation of mixed magnon pairs in a resonator. Experimental examples of combined pumping are given in [92, 93].

4.6 Discussion

a) What is the difference between the linear excitation of magnons and the parametric pumping of magnons?
b) Why does the microwave resonator mode play an important role for the parametric magnon excitation process?
c) Discuss the origin of phase mismatching and nonlinear damping of parametrically pumped magnons.
d) Can a uniform magnetic precession play the role of a resonator? If yes, when?
e) Which is more preferable (and why) for computer simulations of the magnetic dynamics of the magnetic material under coherent alternating magnetic field: (i) solving Landau–Lifshitz–Gilbert equations for individual particles interacting with each other or, (ii) finding collective modes (magnons) in the system and representing magnetic dynamics in terms of dynamics of magnons.

5 Thermodynamic Description of Strongly Excited Magnon System

The above analysis of the magnon parametric excitation was essentially based on the concept of steady energy flow from the pumping field to the thermal bath via the resonator and excited state of resonating magnon pairs. The stability of this flow is supported by the phase mismatching and positive nonlinear damping of magnons.

The theory in which just a narrow spectral resonance region is strongly excited works very well right above the pumping threshold. However this nice and relatively simple model deteriorates when the energy flow through the magnon system increases with the increasing of the microwave pumping. Magnons from the resonance region via four-magnon interaction are transformed to the magnons outside of this region. The deviations of these "secondary" magnons populations from the thermal equilibrium levels become more and more essential. The detailed analysis of essential deviations of magnon populations leads to a cumbersome system of kinetic equations (see [73, 94–96]) and the approach becomes extremely complicated.

In this chapter we shall consider a completely different approach to the description of strongly excited magnon systems. Parametrically pumped magnons evolve to an entirely different state in their finite reservoir when it is practically isolated from other degrees of freedom of the crystal. Collisions between magnons lead to redistribution of the absorbed energy in the Bose gas, and after a certain time τ_{m} (which depends on the magnon density) a steady-state nonequilibrium distribution in this gas is established. The microwave absorption by magnon pairs becomes saturated due to complete mismatching in phase between the forced magnetic oscillations and the applied field. For a real system the approximation of an isolated magnon reservoir is valid, obviously, only for a finite time interval $t \ll \tau_{\mathrm{ph}}$, where τ_{ph} is the characteristic time of the excited magnon relaxation to a phonon system (thermal bath). Thus this kind of evolution of parametrically excited magnons can be observed in experiments that use a pulse technique with pulse length τ: $\tau_{\mathrm{m}} \leq t \ll \tau_{\mathrm{ph}}$. This inequality can be satisfied for a large number of magnetic systems, because magnon–magnon scattering is much more effective (due to exchange interaction) than magnon–phonon processes and in view of the different dependence of τ_{m} and τ_{ph} on the density of the excited magnons.

The saturation effect is easy to understand by using a coordinate frame that "rotates" at a frequency $\omega_{\mathrm{p}}/2$. In this system of coordinates the magnetic Hamiltonian

Nonequilibrium Magnons, First Edition. Vladimir L. Safonov.

of interest to us does not depend explicitly on the time. This Hamiltonian can be reduced by the Bogolyubov canonical transformation to a form that describes a "condensate" of magnon pairs and a weakly nonideal Bose gas of new quasi particles. Turning on the microwave pump field means creation of both the magnon pairs condensate and a definite number of quasi particles whose mutual scattering leads to a thermodynamic equilibrium at a certain effective temperature.

It should be noted that the idea of a thermodynamic description in a rotating frame is very fruitful and popular in the theory of nuclear magnetic resonance (see, for example, [97, 98]). There it was developed for systems with discrete spectra.

Generally speaking, the problem of a strongly driven many-particle Bose system exists and attracts a great deal of interest in different parts of nonlinear physics. Our aim here is to demonstrate a thermodynamic approach that can be applicable to a whole class of systems for which the internal degrees of freedom can be considered as practically isolated from a thermal bath.

5.1 Principal Equations

Our plan consists of: (i) to transform the initial time-dependent Hamiltonian of the system to a new Hamiltonian which no longer depends explicitly on time; (ii) to introduce a thermodynamic description for this new Hamiltonian and find solutions to the problem; (iii) to discuss obtained results with respect to the laboratory system of coordinates. For simplicity we shall consider the process of direct pumping of magnon pairs (without resonator). The role of microwave resonator will be discussed at the end of this chapter.

5.1.1 Hamiltonian

We shall consider the Hamiltonian of a magnon system in the form

$$\mathcal{H} = \mathcal{H}_0 + \mathcal{H}_{\mathrm{p}} + \mathcal{H}_{\mathrm{int}} \,, \tag{5.1}$$

where

$$\frac{\mathcal{H}_0}{\hbar} = \sum_k \omega_k b_k^\dagger b_k \tag{5.2}$$

describes the energy of an ideal magnon gas. Creation and annihilation operators $b_k^\dagger$ and b_k satisfy standard commutation relations for the Bose operators:

$$\begin{aligned} &[b_k, b_q] = [b_k^\dagger, b_q^\dagger] = 0 \,, \\ &[b_k, b_q^\dagger] = \Delta(\boldsymbol{k} - \boldsymbol{q}) \,. \end{aligned} \tag{5.3}$$

The time-dependent Hamiltonian

$$\frac{\mathcal{H}_\mathrm{p}}{\hbar} = \frac{1}{2}\sum_k h V_k \left[b_k b_{-k} \exp(i\omega_\mathrm{p} t) + b_k^\dagger b_{-k}^\dagger \exp(-i\omega_\mathrm{p} t) \right] \tag{5.4}$$

describes the classical microwave pumping field $h(t)$ which excites the magnon pairs via coupling coefficient V_k.

$$\frac{\mathcal{H}_\mathrm{int}}{\hbar} = \frac{1}{2}\sum_{1,2,3,4} \Phi(1,2;3,4) b_1^\dagger b_2^\dagger b_3 b_4 \Delta(\boldsymbol{k}_1 + \boldsymbol{k}_2 - \boldsymbol{k}_3 - \boldsymbol{k}_4) \tag{5.5}$$

is the interaction Hamiltonian, which takes into account the four-magnon scattering. For simplicity, the following notations $\mathbf{1} \equiv \boldsymbol{k}_1, \mathbf{2} \equiv \boldsymbol{k}_2, \mathbf{3} \equiv \boldsymbol{k}_3, \mathbf{4} \equiv \boldsymbol{k}_4$ are used. The scattering amplitude has the symmetry:

$$\begin{aligned} \Phi(1,2;3,4) = \Phi(2,1;3,4) = \Phi(1,2;4,3) = \Phi(2,1;4,3)\ , \\ \Phi(1,2;3,4) = \Phi(3,4;1,2)\ . \end{aligned} \tag{5.6}$$

5.1.2 Unitary Transformation

The time evolution of the system is described by the equation for the density matrix

$$i\hbar \frac{\partial \varrho}{\partial t} = [\mathcal{H}, \varrho]\ . \tag{5.7}$$

Let us introduce a new density matrix ϱ' in the rotating frame applying the unitary transformation

$$\begin{aligned} \varrho' &= U^{-1} \varrho U\ , \\ U &= \exp\left(-\frac{i\omega_\mathrm{p} t}{2} \sum_k b_k^\dagger b_k \right)\ . \end{aligned} \tag{5.8}$$

As a result one obtains

$$i\hbar \frac{\partial \varrho'}{\partial t} = [\mathcal{H}', \varrho']\ , \tag{5.9}$$

where

$$\mathcal{H}' = U \mathcal{H} U^{-1} - \frac{\hbar\omega_\mathrm{p}}{2} \sum_k b_k^\dagger b_k\ . \tag{5.10}$$

Taking into account that

$$\begin{aligned} U b_k U^{-1} &= b_k \exp\left(-i\frac{\omega_\mathrm{p} t}{2} \right)\ , \\ U b_k^\dagger U^{-1} &= b_k^\dagger \exp\left(i\frac{\omega_\mathrm{p} t}{2} \right)\ , \end{aligned} \tag{5.11}$$

we get the new Hamiltonian, which does not have an explicit time dependence:

$$\frac{\mathcal{H}'}{\hbar} = \sum_k \left[\left(\omega_k - \frac{\omega_p}{2} \right) b_k^\dagger b_k + \frac{h V_k}{2} \left(b_k b_{-k} + b_k^\dagger b_{-k}^\dagger \right) \right] + \frac{\mathcal{H}_{int}}{\hbar} . \quad (5.12)$$

For this stationary Hamiltonian one can use a thermodynamic description of the system with the density matrix of the form

$$\varrho' = \frac{\exp(\mathcal{H}'/k_B T_{eff})}{\mathrm{Tr}\left[\exp(\mathcal{H}'/k_B T_{eff})\right]} . \quad (5.13)$$

Here T_{eff} is an effective temperature of the system which, in principle, can differ from the thermal bath temperature.

It is convenient to perform various operator calculations with a thermodynamic density matrix (5.13) if the latter represents a weakly interacting Bose gas. However, as one can see from (5.12), the quadratic part of the new Hamiltonian contains non-diagonal terms. This means that magnons are no longer used to describe normal modes of the system. One can also see that it is not possible to solve the problem of quadratic form diagonalization for all $\boldsymbol{k}$ in (5.12), because in this case the spectrum

$$\Omega_k = \sqrt{\left(\omega_k - \frac{\omega_p}{2} \right)^2 - (h V_k)^2} \quad (5.14)$$

becomes imaginary in the vicinity of resonance point $\omega_k \simeq \omega_p/2$. To overcome this problem it is necessary to take into account higher-order terms.

5.1.3
Bogoliubov Transformation

To find new normal modes of the system (5.12), we shall develop an approach analogous to the Bogolyubov transformation [99], which was developed for a Fermi gas in the theory of superconductivity.

We shall use the transformation of the form:

$$\begin{aligned} b_k &= u_k c_k + v_k c_{-k}^\dagger , \\ b_k^\dagger &= u_k c_k^\dagger + v_k c_{-k} , \end{aligned} \quad (5.15)$$

where

$$\begin{aligned} [c_k, c_q] &= [c_k^\dagger, c_q^\dagger] = 0 , \\ [c_k, c_q^\dagger] &= \Delta(\boldsymbol{k} - \boldsymbol{q}) , \end{aligned} \quad (5.16)$$

and u_k, v_k are unknown coefficients.

Substituting (5.15) into (5.16), one can obtain

$$u_k^2 - v_k^2 = 1 . \quad (5.17)$$

The main idea of (5.15) is to find an explicit form

$$\frac{\mathcal{H}_0'}{\hbar} = \frac{\mathcal{E}_0}{\hbar} + \sum_k \left[\mathcal{A}_k c_k^+ c_k + \frac{\mathcal{B}_k}{2} (c_k c_{-k} + c_k^\dagger c_{-k}^\dagger) \right] + \frac{\mathcal{H}_{\text{int}}'}{\hbar} \tag{5.18}$$

using four-magnon terms and applying the condition

$$\mathcal{B}_k = 0 \,, \tag{5.19}$$

which is the second equation for u_k and v_k coefficients.

In this case the quadratic form becomes a diagonal one and $\mathcal{A}_k$ describes the spectrum Ω_k of new quasi particles. Since these quasi particles contain oscillations of the magnetic system induced by the external pumping field, we shall call them *inductons* to distinguish them from the bare magnons. $\mathcal{H}_{\text{int}}'$ in (5.18) denotes the normally ordered Hamiltonian of four-inducton interactions.

The calculation is straightforward, for example,

$$\begin{aligned} b_k^\dagger b_k &= \left(u_k c_k^\dagger + v_k c_{-k} \right) \left(u_k c_k + v_k c_{-k}^\dagger \right) \\ &= u_k^2 c_k^\dagger c_k + v_k^2 c_{-k} c_{-k}^\dagger + u_k v_k \left(c_k^\dagger c_{-k}^\dagger + c_{-k} c_k \right) \\ &= v_k^2 + \left(u_k^2 + v_k^2 \right) c_k^\dagger c_k + u_k v_k \left(c_k^\dagger c_{-k}^\dagger + c_{-k} c_k \right) \end{aligned} \tag{5.20}$$

and

$$\begin{aligned} b_k b_{-k} &= \left(u_k c_k + v_k c_{-k}^\dagger \right) \left(u_k c_{-k} + v_k c_k^\dagger \right) \\ &= u_k v_k \left(c_k c_k^\dagger + c_{-k}^\dagger c_{-k} \right) + u_k^2 c_k c_{-k} + v_k^2 c_{-k}^\dagger c_k^\dagger \\ &= u_k v_k + u_k v_k \left(c_k^\dagger c_k + c_{-k}^\dagger c_{-k} \right) + u_k^2 c_k c_{-k} + v_k^2 c_{-k}^\dagger c_k^\dagger \,. \end{aligned} \tag{5.21}$$

Quite similar calculations are used for the four-quasi particle Hamiltonian. As a result, we obtain

$$\begin{aligned} \mathcal{A}_k = &\left(\omega_k - \frac{\omega_\text{p}}{2} + 2 \sum_q \mathcal{T}_{kq} v_q^2 \right) \left(2 v_k^2 + 1 \right) \\ &+ 2 \left(h V_k + \sum_q \mathcal{S}_{kq} u_q v_q \right) u_k v_k \,, \end{aligned} \tag{5.22}$$

and

$$\begin{aligned} \mathcal{B}_k = &\, 2 \left(\omega_k - \frac{\omega_\text{p}}{2} + 2 \sum_q \mathcal{T}_{kq} v_q^2 \right) u_k v_k \\ &+ \left(h V_k + \sum_q \mathcal{S}_{kq} u_q v_q \right) \left(2 v_k^2 + 1 \right) \,, \end{aligned} \tag{5.23}$$

where

$$\mathcal{T}_{kq} \equiv \Phi(k, q; k, q) \,, \quad \mathcal{S}_{kq} \equiv \Phi(k. - k; q, -q) \,. \tag{5.24}$$

We can write an explicit form of the four-inducton Hamiltonian $\mathcal{H}'_{\text{int}}$ as

$$\begin{aligned}
\frac{\mathcal{H}'_{\text{int}}}{\hbar} &= \frac{1}{2} \sum_{1,2,3,4} \tilde{\Phi}_2(1,2;3,4) c_1^\dagger c_2^\dagger c_3 c_4 \Delta(k_1 + k_2 - k_3 - k_4) \\
&+ \frac{1}{2} \sum_{1,2,3,4} \left[\tilde{\Phi}_1(1;2,3,4) c_1^\dagger c_2 c_3 c_4 + \text{h.c.} \right] (k_1 - k_2 - k_3 - k_4) \\
&+ \frac{1}{2} \sum_{1,2,3,4} \left[\tilde{\Phi}_0(1,2,3,4) b_1^\dagger b_2^\dagger b_3 b_4 + \text{h.c.} \right] \Delta(k_1 + k_2 - k_3 - k_4) \,,
\end{aligned} \tag{5.25}$$

where

$$\begin{aligned}
\tilde{\Phi}_2(1,2;3,4) &= \Phi(1,2;3,4) u_1 u_2 u_3 u_4 + \Phi(3,4;1,2) v_1 v_2 v_3 v_4 \\
&\quad + \Phi(-3,2;-1,4) u_1 v_2 v_3 u_4 + \Phi(1,-3;-2,4) u_1 v_2 v_3 u_4 \\
&\quad + \Phi(-4,2;3,-1) v_1 u_1 u_1 u_1 + \Phi(1,-4;3,-2) u_1 u_1 u_1 u_1 \,, \\
\tilde{\Phi}_1(1;2,3,4) &= \Phi(2,1;3,4) u_1 v_2 u_3 u_4 + \Phi(1,2;3,4) u_1 v_2 u_3 u_4 \\
&\quad + \Phi(-3,-2;-1,4) v_1 v_2 v_3 u_4 + \Phi(-4,-3;-2,-1) u_1 v_2 u_3 v_4 \,, \\
\tilde{\Phi}_0(1,2,3,4) &= \Phi(-1,-2;3,4) v_1 v_2 u_3 u_4
\end{aligned}$$

are the amplitudes of inducton interactions, which are given for simplicity in the nonsymmetric forms.

5.1.4
Effective Temperature $T_{\text{eff}} = 0$

Consider first the case of zero effective temperature, when the occupation numbers of inductons are equal to zero. This means that the "vacuum" state of the system is described by the vacuum-state *bra* and *ket* vectors:

$$c_k|0\rangle = 0 \quad \text{and} \quad \langle 0|c_k^\dagger = 0 \,. \tag{5.26}$$

The energy of the vacuum state is equal to

$$\langle 0|\mathcal{H}'|0\rangle = \mathcal{E}_0 = \frac{\hbar}{2} \sum_k \left[\left(\omega_k - \frac{\omega_{\text{p}}}{2} + E_k \right) N_k + (h V_k + \Delta_k) \sigma_k \right] . \tag{5.27}$$

Here for convenience we use the following notations:

$$\begin{aligned}
N_k &\equiv \langle 0|b_k^\dagger b_k|0\rangle = v_k^2 \,, \\
\sigma_k &\equiv \langle 0|b_k b_{-k}|0\rangle = u_k v_k \,,
\end{aligned} \tag{5.28}$$

and

$$E_k = \omega_k - \frac{\omega_p}{2} + 2\sum_q \mathcal{T}_{kq} N_q ,$$

$$\Delta_k = h V_k + \sum_q \mathcal{S}_{kq} \sigma_q . \tag{5.29}$$

Equation (5.17) in these notations takes the form

$$\sigma_k^2 = N_k^2 + N_k . \tag{5.30}$$

Equation (5.19) can be represented as

$$E_k \sigma_k + \left(N_k + \frac{1}{2} \right) \Delta_k = 0 . \tag{5.31}$$

This equation can also be obtained by minimizing (5.27) with the constraint of (5.30). The functions N_k and σ_k describe the distribution of the excited Bose gas calculated from the vacuum state of the Hamiltonian of inductons

$$\mathcal{H}' = \mathcal{E}_0 + \sum_k \hbar \Omega_k c_k^\dagger c_k + \mathcal{H}'_{\text{int}} . \tag{5.32}$$

The spectrum of inductons has the form:

$$\Omega_k = \sqrt{E_k^2 - \Delta_k^2 \text{sign}(\Delta_k)} . \tag{5.33}$$

It is easy to obtain the condition under which this spectrum is real:

$$E_k^2 - \Delta_k^2 = \frac{1}{4} \frac{\Delta_k^2}{N_k^2 + N_k} \geq 0 . \tag{5.34}$$

It can be seen that it is valid for solutions of physical interest, with $N_k > 0$. It should be noted that the change of the sign of the spectrum (5.33) does not effect the dynamic part of the problem and for the thermodynamic description it will be compensated by the effective chemical potential μ.

5.1.5
Effective Temperature $T_{\text{eff}} \neq 0$

The effective temperature of the inducton gas depends in general on the initial conditions. It may happen that no vacuum state with $T_{\text{eff}} = 0$ is realized under any physical assumption. It is necessary therefore to find a generalization for the case of $T_{\text{eff}} \neq 0$ for (5.17) and (5.19) for the coefficients of Bogolyubov transformation that reduces the Hamiltonian (5.12) to normal modes, inductons. Such a generalization is possible in the framework of the self-consistent field approximation if the inducton occupation numbers are relatively small. We can introduce the thermodynamic density matrix of the ideal inducton gas in the form

$$\varrho_0 = \frac{e^{\mathcal{Q}}}{\mathrm{Tr}(e^{\mathcal{Q}})}\,,$$
$$\mathcal{Q} \equiv -\frac{1}{k_\mathrm{B} T_\mathrm{eff}} \sum_k (\hbar\Omega_k - \mu) c_k^+ c_k\,, \tag{5.35}$$

where μ is the effective chemical potential.

Let us define the following averages

$$N_k = \left\langle b_k^\dagger b_k \right\rangle_0 = \mathrm{Tr}\left(b_k^\dagger b_k \varrho_0\right)$$

and

$$\sigma_k = \langle b_k b_{-k} \rangle_0 = \mathrm{Tr}(b_k b_{-k} \varrho_0)$$

over the density matrix ϱ_0 with allowance for relations (5.15); then

$$N_k = u_k^2 n_k + (n_k + 1) v_k^2\,,$$
$$\sigma_k = u_k v_k (1 + 2n_k)\,, \tag{5.36}$$

where $n_k \equiv \langle c_k^+ c_k \rangle_0$ describes the occupation number of inductons

$$n_k = \left[\exp\left(\frac{\hbar\Omega_k - \mu}{k_\mathrm{B} T_\mathrm{eff}}\right) - 1\right]^{-1}. \tag{5.37}$$

We can rewrite (5.36) in the form

$$v_k^2 = \frac{N_k - n_k}{2n_k + 1}\,,$$
$$u_k v_k = \frac{\sigma_k}{2n_k + 1}\,. \tag{5.38}$$

Equation (5.17) in the N_k and σ_k notations takes the form

$$\sigma_k^2 = \left(N_k + \frac{1}{2}\right)^2 - \left(n_k + \frac{1}{2}\right)^2. \tag{5.39}$$

The energy of vacuum state can be written as

$$\langle \mathcal{H}' \rangle = \frac{\hbar}{2} \sum_k \left[\left(\omega_k - \frac{\omega_\mathrm{P}}{2} + E_k\right) N_k + (h V_k + \Delta_k)\sigma_k\right]. \tag{5.40}$$

Here we have used the self-consistent approximation for the four-magnon averages

$$\left\langle b_1^\dagger b_2^\dagger b_3 b_4 \right\rangle_0 = [\Delta(1-3)\Delta(2-4) + \Delta(1-4)\Delta(2-3)] N_1 N_2$$
$$+ \Delta(1+2)\Delta(3+4)\sigma_1\sigma_3\,. \tag{5.41}$$

From the condition of the minimum of $\langle \mathcal{H}' \rangle$ in (5.40) with a specified number of n_k we can obtain the second equation for variables N_k and σ_k:

$$E_k \sigma_k + \left(N_k + \frac{1}{2} \right) \Delta_k = 0 . \tag{5.42}$$

Here

$$E_k = \omega_k - \frac{\omega_p}{2} + 2 \sum_q \mathcal{T}_{kq} N_q ,$$

$$\Delta_k = h V_k + \sum_q \mathcal{S}_{kq} \sigma_q . \tag{5.43}$$

5.1.5.1 Maximum of Entropy

Now the only undetermined parameter is the effective chemical potential, which enters explicitly in expression (5.37). We can find it in the ideal gas approximation from the condition of maximum entropy (e.g., [100]):

$$\begin{aligned} S &= \sum_k [(1 + n_k) \ln (1 + n_k) - n_k \ln (n_k)] \\ &= \sum_k \left[\frac{\hbar \Omega_k - \mu}{k_B T_{\text{eff}}} (1 + n_k) + \ln(n_k) \right] . \end{aligned} \tag{5.44}$$

One can see that maximum entropy occurs at $n_k \to \infty$ and is satisfied if

$$\mu = \min(\hbar \Omega_k) \quad \text{at} \quad T_{\text{eff}} > 0 \tag{5.45}$$

This equation for an ideal Bose gas corresponds to the Bose–Einstein condensation condition. Thus, applying strong microwave pumping to a magnetic system, we can expect both the excitation of magnon pairs in the vicinity of parametric resonance and infinitely large accumulation of inductons (magnons) and their Bose–Einstein condensation at the bottom of the spectrum. This fact was theoretically predicted in [101, 102]. At that time the theory (see also [103]) qualitatively explained the experimentally observed microwave irradiation (which corresponds to magnon accumulation) from the bottom of the spectrum in different parametrically excited magnetic systems: (i) electronic magnons in YIG, at room temperature [104, 105], and (ii) nuclear magnons in $CsMnF_3$, at liquid helium temperatures [106]. However, the real discovery of Bose–Einstein condensation of magnons occurred two and half decades later [107].

A coherent state, which corresponds to Bose–Einstein condensate of inductons (magnons) is a macroscopic rotation of magnetization, which will irradiate coherent microwave and its harmonics. So far as the frequency of this motion, bottom frequency of the spectrum, does not depend on the pump frequency, we have an example of controlled microwave energy convertor which can absorb microwaves with different frequencies and transform this energy into bottom frequency microwaves. The latter can be controlled by the magnetic field, and so on. A conversion of a broadband microwave energy accumulated by a system of strongly excited magnons into a monochromatic microwave was demonstrated in [108].

It is interesting to note that formally maximum entropy (5.44) can occur if

$$\mu = \max(\hbar \Omega_k) \quad \text{at} \quad T_{\text{eff}} < 0 \,. \tag{5.46}$$

Below we shall confine ourselves to the case of positive effective temperatures.

5.2
Exact Solutions

As follows from (5.42) and (5.43), in $\boldsymbol{k}$-space there is a resonance surface $\boldsymbol{k}_0 \in \boldsymbol{K}_0$, defined by

$$E_{k_0} = \omega_{k_0} - \frac{\omega_p}{2} + 2\sum_q T_{k_0 q} N_q = 0 \,, \tag{5.47}$$

and

$$\Delta_{k_0} = h V_{k_0} + \sum_q S_{k_0 q} \sigma_q = 0 \,. \tag{5.48}$$

Outside of this surface (5.39) and (5.42) can be represented in the form convenient for iterations:

$$\sigma_k = -\left(N_k + \frac{1}{2}\right) \frac{\Delta_k}{E_k} \,, \tag{5.49}$$

$$N_k = \frac{1}{2}\left(\frac{2n_k + 1}{\sqrt{1 - (\Delta_k / E_k)^2}} - 1\right) \,. \tag{5.50}$$

The expressions for E_k and Δ_k can be written, with allowance for (5.47) and (5.48), in the form

$$E_k = \omega_k - \omega_{k_0} + 2\sum_q \left(T_{kq} - T_{k_0 q}\right) N_q \,,$$

$$\Delta_k = h\left(V_k - V_{k_0}\right) + \sum_q \left(S_{kq} - S_{k_0 q}\right) \sigma_q \,. \tag{5.51}$$

Equations (5.47)–(5.50) can be solved analytically if the matrix elements T_{kq} and S_{kq} have a definite symmetry:

$$T_{kq} = T(\boldsymbol{k} - \boldsymbol{q}) \,, \quad S_{kq} = S(\boldsymbol{k} - \boldsymbol{q}) \,, \tag{5.52}$$

or

$$T_{kq} = T(\boldsymbol{k}) T(\boldsymbol{q}) \,, \quad S_{kq} = S(\boldsymbol{k}) S(\boldsymbol{q}) \,. \tag{5.53}$$

Here, for simplicity we shall consider the case when

$$T_{kq} = 0 \,, \quad S_{kq} = S \,, \quad V_k = V \tag{5.54}$$

for all $\boldsymbol{k}$. Then from (5.51) it follows that

$$E_k = \omega_k - \omega_{k_0} , \tag{5.55}$$

$$\Delta_k = 0 . \tag{5.56}$$

Therefore,

$$\sigma_k = 0 , \quad N_k = n_k \tag{5.57}$$

for all $\boldsymbol{k}$ outside the resonance surface.

On the resonance surface we have

$$\omega_{k_0} - \frac{\omega_p}{2} = 0 , \tag{5.58}$$

$$hV + \mathcal{S} \sum_{k \in K_0} \sigma_k = 0 . \tag{5.59}$$

The solution on the resonance surface can be represented as

$$\sigma_{k_0} = -\frac{hV}{\mathcal{S}\mathcal{M}} ,$$

$$N_{k_0} = \sqrt{\left(\frac{hV}{\mathcal{S}\mathcal{M}}\right)^2 + \left(n_{k_0} + \frac{1}{2}\right)^2} - \frac{1}{2} , \tag{5.60}$$

where

$$\mathcal{M} \equiv \sum_{k \in K_0} 1 \simeq \frac{1}{\pi} (Lk_0)^2 ,$$

is the "measure" of the resonance surface and L is a linear size of the crystal.

Thus we obtain an exact solution of the nonequilibrium many-body problem using the thermodynamic method. It has been shown that the steady-state energy of parametrically excited Bose particles insulated from a thermostat can be represented by two terms. The first is a narrow peak in the vicinity of parametric resonance, which corresponds to the coherent with the pump field "condensate" of magnon pairs, while the second describes an equilibrium inducton gas having a certain effective temperature and the effective chemical potential equal to the bottom of inducton spectrum (the condition of Bose–Einstein condensation). Thus, the population of inductons rapidly increases with approaching the bottom of the spectrum. Such a distribution function, which includes both terms, is shown schematically in Figure 5.1b. Figure 5.1a shows the experimental result [109] with a peak of parametrically pumped magnons and a growth of population of the excited quasi particles to the bottom of the spectrum. There is some specificity of the magnetic system in [109] (thin magnetic film), but, in general, we can see a qualitative agreement between the theoretical prediction and experimental data.

Problem 5.1. Generalize the exact solution for the case of $\mathcal{T}_{kq} = \mathcal{T} \neq 0$.

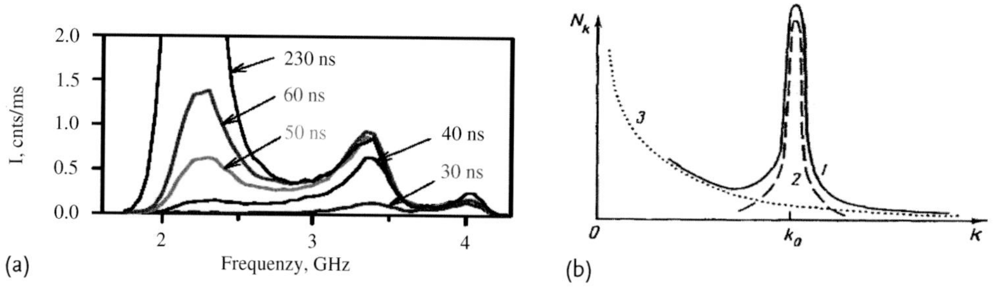

Figure 5.1 (a) The BLS spectra were measured after the pumping is turned on at delays with respect to the start of the pumping pulse as indicated (from Figure 2c in [109]). (b) The magnon distribution function (solid line) is drawn (from Figure 1 in [102]), consisting of the coherent distribution (pair condensate, dashed line) and the incoherent thermodynamic distribution (dotted line).

5.2.1
The Effective Temperature

The effective temperature T_{eff} reached by an inducton gas after the pump field is switched on $(0 \to h V_k)$ can be calculated analytically in two important cases: (i) instantaneous switching and (ii) adiabatic switching.

5.2.1.1 Instantaneous Switching

When the pump field is switched on instantaneously, there is not enough time to change the initial density matrix $\varrho_{\text{m}}(T)$ of the magnon system that describes the equilibrium magnon gas at the temperature T. This allows us to calculate the energy U_{m} of the initially nonequilibrium state of the inducton gas in the rotating coordinate frame, using the following functions

$$N_k^{(\text{m})} = \text{Tr}\left(b_k^\dagger b_k \varrho_{\text{m}}\right) = \frac{1}{\exp(\hbar\omega_k / k_{\text{B}} T) - 1} \tag{5.61}$$

and

$$\sigma_k^{(\text{m})} = \text{Tr}(b_k b_{-k} \varrho_{\text{m}}) = 0 \,. \tag{5.62}$$

Taking into account the change of the chemical potential, we get

$$U_{\text{m}} = \sum_k \hbar\omega_k N_k^{(\text{m})} \,. \tag{5.63}$$

Since the Hamiltonian does not depend explicitly on time, the energy is conserved. For thermodynamic equilibrium, the expression for the inducton gas energy is

$$U = \langle \mathcal{H}' \rangle - \mu \sum_k n_k \,, \tag{5.64}$$

where $\langle \mathcal{H}' \rangle$ is defined by (5.40), while the second term is due to the indeterminate number of inductons. Thus, the equation

$$U_m(T) = U(T_{\text{eff}}) \tag{5.65}$$

yields an explicit dependence T_{eff} on T. For example, for the exact solution considered above, when $\mathcal{T}_{kq} = 0$, $\mathcal{S}_{kq} = \mathcal{S}$ and $V_k = V$ this equation takes the form

$$\sum_k \frac{\hbar\omega_k}{\exp(\hbar\omega_k / k_B T) - 1} = -\frac{(hV)^2}{2\mathcal{S}} + \frac{\hbar}{2}\sum_k \frac{\omega_k - \omega_0}{\exp[\hbar(\omega_k - \omega_0)/k_B T_{\text{eff}}] - 1} . \tag{5.66}$$

As it follows from this formula, the sign of $\mathcal{S}$ determines the character of the dependence of the temperature T_{eff} on the microwave amplitude h. For the case of $\mathcal{S} > 0$ an increase of h causes T_{eff} to increase, and the inducton gas cools down in the opposite case. This cooling continues until zero temperature at a given microwave power h_0. With further increase of h it follows from (5.66) that only negative effective temperatures are possible. In this case the parametrically excited system seems not to reach a stationary state and perform a finite motion with constant energy.

5.2.1.2 Adiabatic Switching

If the pump field is turned on adiabatically the entropy $\mathrm{S}[T_{\text{eff}}(t)]$ of the system is constant. It is assumed that a new quasi particle system temperature sets after each small change of the pump amplitude h. The chemical potential is determined then from the maximum entropy. The equation for the effective temperature T_{eff} as a function of T is

$$\mathrm{S}(\hbar\Omega_k - \mu, T_{\text{eff}}) = \mathrm{S}(\hbar\omega_k, T) , \tag{5.67}$$

where

$$\mathrm{S}(\hbar\omega_k, T) = \sum_k \left\{ \frac{\hbar\omega_k}{k_B T}\left[\exp\left(\frac{\hbar\omega_k}{k_B T}\right) - 1\right]^{-1} - \log\left[1 - \exp\left(-\frac{\hbar\omega_k}{k_B T}\right)\right]\right\} , \tag{5.68}$$

and the expression for $\mathrm{S}(\hbar\Omega_k - \mu, T_{\text{eff}})$ is similar. It is clear that in the general case $T_{\text{eff}} < T$, but each specific case must be separately computed.

5.2.1.3 Thermodynamic Stability

The question of thermodynamic stability of the equilibrium state of inducton Bose gas reduces to satisfaction of the following thermodynamic criteria [100]: (i) the heat capacity C_V at constant volume must be positive, and (ii) the isotropic compression increases the pressure.

An explicit form of condition (i) can be obtained by differentiating energy U (5.64) with respect to T_{eff}:

$$\frac{dU}{dT_{\text{eff}}} > 0 \,. \tag{5.69}$$

For the above example of exact solution we have

$$\sum_k (\omega_k - \omega_0)^2 \exp\left[\frac{\hbar(\omega_k - \omega_0)}{k_B T_{\text{eff}}}\right] \left\{\exp\left[\frac{\hbar(\omega_k - \omega_0)}{k_B T_{\text{eff}}}\right] - 1\right\}^{-2} > 0 \,, \tag{5.70}$$

which is obviously always valid.

The second condition is easy to verify by using the equation of state of an ideal Bose gas:

$$P V_s = -k_B T_{\text{eff}} \sum_k \ln\left[1 - \exp\left(\frac{\mu - \hbar\Omega_k}{k_B T_{\text{eff}}}\right)\right] . \tag{5.71}$$

Since the inducton spectrum Ω_k is independent from the pressure P and system volume V_s, we have $P V_s = \text{const.}$ for a given temperature. Thus the isotropic compression increases the pressure and the criterion is true in this case.

5.2.2
Collective Oscillations

If the gas of inductons is thermodynamically stable, small density perturbations relax to an equilibrium within a characteristic time τ determined by the four-quasi particle part $\mathcal{H}'_{\text{int}}$ (5.32). The relaxation process due to the phase relations is oscillatory and its characteristic frequency Ω is referred to as the system collective oscillation frequency. We shall consider the approximation $\Omega \gg \tau^{-1}$.

Let us write dynamic equations for the wave pairs in term of N_k and σ_k. These equations can be obtained from the Heisenberg equations for $b_k^+ b_k$ and $b_k b_{-k}$ in the self-consistent field approximation by a Bogolyubov method [99]. We omit this simple procedure and write the final result:

$$\begin{aligned}
\frac{i}{2}\frac{d}{dt}\sigma_k &= E_k \sigma_k + \left(N_k + \frac{1}{2}\right)\Delta_k \,, \\
-\frac{i}{2}\frac{d}{dt}\sigma_k^* &= E_k \sigma_k^* + \left(N_k + \frac{1}{2}\right)\Delta_k^* \,, \\
i\frac{d}{dt} N_k &= \Delta \sigma_k^* - \Delta_k^* \sigma_k \,.
\end{aligned} \tag{5.72}$$

Combining these system of equations, we can obtain

$$\frac{d}{dt}\left[|\sigma_k|^2 - \left(N_k + \frac{1}{2}\right)^2\right] = 0 \,. \tag{5.73}$$

Taking into account (5.39), one has an integral of motion

$$|\sigma_k|^2 = \left(N_k + \frac{1}{2}\right)^2 - \left(n_k + \frac{1}{2}\right)^2 . \tag{5.74}$$

Now let us investigate the spectrum of small deviations

$$\delta N_k(t) = \delta N_{k\lambda} e^{\lambda t} , \tag{5.75}$$

$$\delta \sigma_k(t) = \delta \sigma_{k\lambda} e^{\lambda t} \tag{5.76}$$

from equilibrium values N_k, and σ_k.

Linearization of (5.72) and (5.74) yields

$$i\lambda \delta \sigma_{k\lambda} = 2E_k \delta \sigma_{k\lambda} + 2\sigma_k \delta E_{k\lambda} + 2\Delta_k \delta N_{k\lambda} + (2N_k + 1)\delta \Delta_{k\lambda} , \tag{5.77}$$

$$i\lambda \delta N_{k\lambda} = \Delta_k(\delta \sigma^*_{k\lambda} - \delta \sigma_{k\lambda}) + \sigma_k(\delta \Delta_{k\lambda} - \delta \Delta^*_{k\lambda}) , \tag{5.78}$$

$$\sigma_k \left(\delta \sigma_{k\lambda} + \delta \sigma^*_{k\lambda}\right) = (2N_k + 1)\delta N_{k\lambda} , \tag{5.79}$$

where

$$\delta E_{k\lambda} = 2\sum_q \mathcal{T}_{kq} \delta N_{q\lambda} , \tag{5.80}$$

$$\delta \Delta_{k\lambda} = \sum_q \mathcal{S}_{kq} \delta \sigma_{q\lambda} . \tag{5.81}$$

Equations (5.77)–(5.79) cannot be solved analytically in the general case. For simplicity, we shall consider the case of constant coefficients $\mathcal{T}_{kq} = \mathcal{T}$, $\mathcal{S}_{kq} = \mathcal{S}$, $V_k = V$, when the solution for σ_k is singular and has spherical symmetry.

Outside the resonance surface $k \neq k_0$ we have then, from (5.77) and (5.78),

$$\delta N_{k\lambda} = 0 , \tag{5.82}$$

$$\delta \sigma_{k\lambda} = \frac{2n_k + 1}{i\lambda - 2(\omega_k - \omega_{k_0})} \mathcal{S} \sum_q \delta \sigma_{q\lambda} . \tag{5.83}$$

Summing both sides of (5.83) over $\boldsymbol{k}$, we get

$$\sum_k \delta \sigma_{k\lambda} = \mathcal{M}\Lambda(\lambda) \delta \sigma_{k_0\lambda} , \tag{5.84}$$

where

$$\Lambda(\lambda) \equiv \left[1 - \mathcal{S} \sum_{k \notin K_0} \frac{2n_k + 1}{i\lambda - 2(\omega_k - \omega_{k_0})}\right]^{-1} . \tag{5.85}$$

With this relation simple transformations of (5.77) and (5.78) lead to the equations for small deviations on the resonance surface:

$$\left[i\lambda + \mathcal{M}\mathcal{S}\Lambda^*(2N_{k_0}+1)\right]\delta N_{k_0\lambda} = 2\mathcal{M}\mathcal{S}\sigma_{k_0}\,\mathrm{Re}(\Lambda)\delta\sigma_{k_0\lambda}\,, \tag{5.86}$$

$$\left[i\lambda - \mathcal{M}\mathcal{S}\Lambda(2N_{k_0}+1)\right]\delta\sigma_{k_0\lambda} = 4\mathcal{M}\mathcal{T}\sigma_{k_0}\delta N_{k_0\lambda}\,. \tag{5.87}$$

From the condition that these systems of equations have nontrivial solutions we obtain an equation for the collective oscillation spectrum

$$\lambda^2 - 4\mathcal{M}\mathcal{S}\left(N_{k_0}+\frac{1}{2}\right)\mathrm{Im}(\Lambda)\lambda + 4\left[\mathcal{M}\mathcal{S}\left(N_{k_0}+\frac{1}{2}\right)\right]^2|\Lambda|^2 + 8\mathcal{M}^2\mathcal{T}\mathcal{S}\,\mathrm{Re}(\Lambda)\sigma_{k_0}^2 = 0\,. \tag{5.88}$$

In the single-mode approximation $\Lambda = 1$ we obtain the following solution

$$\Omega = |\,\mathrm{Im}\,\lambda| = 2\left[(hV)^2\left(1+\frac{2\mathcal{T}}{\mathcal{S}}\right) + (\mathcal{M}\mathcal{S})^2\left(n_{k_0}+\frac{1}{2}\right)^2\right]^{1/2}. \tag{5.89}$$

Equation (5.88) also yields a criterion of stability for the stationary state. Stability is lost in a Hamiltonian system only when λ passes through zero. Thus the stationary state is stable if

$$\left(\frac{hV}{\mathcal{M}\mathcal{S}^2}\right)^2\mathcal{S}\left(\frac{2\mathcal{T}}{\Lambda(0)}+\mathcal{S}\right) + \left(n_{k_0}+\frac{1}{2}\right)^2 > 0\,. \tag{5.90}$$

5.3
Magnon Pumping in a Resonator

Now we consider the pumping of a magnetic system in a microwave resonator. In this case the Hamiltonian of magnon pumping can be written as:

$$\frac{\mathcal{H}_p}{\hbar} = F\left[R^\dagger\exp(-i\omega_p t) + R\exp(i\omega_p t)\right] + \omega_R R^\dagger R + \frac{i}{2}\sum_k G_k\left(Rb_k^\dagger b_{-k}^\dagger - R^\dagger b_k b_{-k}\right). \tag{5.91}$$

The unitary transformation

$$\varrho' = U^{-1}\varrho U\,,\quad U = \exp\left[-i\omega_p t\left(R^*R + \frac{1}{2}\sum_k b_k^\dagger b_k\right)\right] \tag{5.92}$$

gives the following time-independent Hamiltonian

$$\frac{\mathcal{H}'}{\hbar} = F(R^\dagger + R) + (\omega_R - \omega_p) R^\dagger R + \sum_k \left[\left(\omega_k - \frac{\omega_p}{2} \right) b_k^\dagger b_k + \frac{i G_k}{2} \left(R b_k^\dagger b_{-k}^\dagger - R^\dagger b_k b_{-k} \right) \right] + \frac{\mathcal{H}_{\text{int}}}{\hbar} . \tag{5.93}$$

From the dynamic equation of motion for the resonator variables at $\omega_p = \omega_R$ one can obtain the following stationary solution:

$$R = \frac{1}{i\Gamma} \left(F - \frac{i}{2} \sum_q G_q b_q b_{-q} \right) ,$$

$$R^\dagger = -\frac{1}{i\Gamma} \left(F + \frac{i}{2} \sum_q G_q b_q^\dagger b_{-q}^\dagger \right) . \tag{5.94}$$

Substituting this solution into (5.93), we get

$$\frac{\mathcal{H}'}{\hbar} = \sum_k \left[\left(\omega_k - \frac{\omega_p}{2} \right) b_k^\dagger b_k + \frac{F G_k}{2\Gamma} \left(b_k^\dagger b_{-k}^\dagger + b_k b_{-k} \right) \right] + \frac{\mathcal{H}_{\text{int}}}{\hbar} ,$$

which exactly corresponds to (5.12) with $h V_k = F G_k / \Gamma$. Thus, in the saturated state the resonator gives high amplitude of the microwave field and does not play any other specific role.

Problem 5.2. Develop the thermodynamic theory [110] of magnon excitation via the uniform precession pumping

$$\frac{\mathcal{H}_p}{\hbar} = h V \left[b_0^\dagger \exp\left(-i\omega_p t\right) + b_0 \exp(i\omega_p t) \right] \tag{5.95}$$

using the unitary transformation

$$\varrho' = U^{-1} \varrho U ,$$

$$U = \exp\left(-i\omega_p t \sum_k b_k^\dagger b_k \right) . \tag{5.96}$$

5.4 Discussion

a) Why and how can thermodynamic methods be applied to a strongly excited Bose system?
b) Give several examples of rotating frames.
c) Is it possible to develop a thermodynamic theory for mixed magnon pairs?
d) Is it possible to excite collective oscillations in the saturated magnon system?

6
Bose–Einstein Condensation of Quasi Equilibrium Magnons

6.1
Bose Gas of Magnons

As it follows from classical physics, the gas of particles with attractive interaction undergoes phase transitions to liquid and then to solid states during lowering of temperature. The finite temperatures of these phase transitions are defined by the attraction strength. This simple picture is in agreement with our everyday intuition: slowly moving attractive particles have to stick together. In the case of an ideal gas, in which there is no interaction between particles, the classical theory predicts a gaseous state down to absolute zero temperature.

However, completely different predictions for ideal gases follow from the quantum statistical theory, in which particles are divided into two quite different classes: fermions and bosons. Fermionic properties of conduction electrons play an extremely important role in the theory of metals and semiconductors (e.g., [100, 111]). In this book, however, we focus on bosons.

As it follows from definition, magnons, the quanta of spin oscillations in $\boldsymbol{k}$-space, obey Bose–Einstein statistics. One of the most remarkable properties of quantum statistics is that the ideal gas of Bose particles can exhibit the phenomenon of Bose–Einstein condensation (see, for example, [112]) at low temperatures, one of the most intriguing and fascinating phase transition in physics.

Does the fact that magnons are bosons automatically mean that there are conditions when Bose–Einstein condensation occurs for magnon gas? In this chapter we shall analyze this question in detail and discuss related problems.

6.1.1
Ideal Bose Gas

Let us consider a general form of the boson population:

$$n_k = \left[\exp\left(\frac{\hbar\omega_k - \mu}{k_B T}\right) - 1\right]^{-1} , \qquad (6.1)$$

where the chemical potential μ is always less than the energy $\hbar\omega_k$ in order to have positive and finite n_k. The chemical potential is defined as a solution of the integral

Nonequilibrium Magnons, First Edition. Vladimir L. Safonov.

equation

$$N = \sum_{k} n_k = \sum_{k} \left[\exp\left(\frac{\hbar\omega_k - \mu}{k_B T} \right) - 1 \right]^{-1} , \tag{6.2}$$

where $N =$ const. is the total number of bosons. From this formula follows the condition of Bose–Einstein condensation $\mu = 0$ for real particles with mass m and energy $\hbar\omega_k = (\hbar k)^2/2m$

$$\frac{N}{V_s} = \int_0^\infty \frac{4\pi k^2 dk}{(2\pi)^3} \left[\exp\left(\frac{(\hbar k)^2}{2mk_B T} - \frac{\mu}{k_B T} \right) - 1 \right]^{-1}$$
$$= \frac{(2mk_B T)^{3/2}}{8\hbar^3 \pi^{3/2}} \mathrm{Li}_{3/2} \left[\exp\left(\frac{\mu}{k_B T} \right) \right] . \tag{6.3}$$

Here V_s is the volume of the system and

$$\mathrm{Li}_\nu(z) \equiv \sum_{n=1}^{\infty} \frac{z^n}{n^\nu} \tag{6.4}$$

is the PolyLog function.

According to (6.3), the negative chemical potential increases with decreasing temperature and approaches zero. Taking into account that

$$\lim_{\mu \to -0} \mathrm{Li}_{3/2} \left[\exp\left(\frac{\mu}{k_B T} \right) \right] = \zeta\left(\frac{3}{2} \right) \simeq 2.61 , \tag{6.5}$$

where $\zeta(\cdot)$ is the zeta-function, we can write the relation

$$T_{\mathrm{BEC}} = 3.32 \frac{\hbar^2}{m k_B} \left(\frac{N}{V_s} \right)^{2/3} \tag{6.6}$$

for the critical temperature T_{BEC} as the function of the particles density N/V_s.

Bose–Einstein condensation was discovered experimentally in dilute atomic gases (see, for example [113–115]). The main fundamental interest to the phenomenon is that the coherent state (in $\boldsymbol{k}$-space) appears *without attraction.*

There is a principal difference between magnons as quasi particles and real Bose particles. The number of real particles is not changed ($N =$ const.) when the temperature decreases. But the number of magnons decreases ($N \to 0$ at $T \to 0$) with decreasing temperature. There are no spin excitations, i.e., no magnons ($N = 0$) at $T = 0$ K. This is why the chemical potential of magnons in the thermodynamic equilibrium is always equal to zero ($\mu = 0$) and the total magnon number is defined by

$$N = \sum_{k} \left[\exp\left(\frac{\hbar\omega_k}{k_B T} \right) - 1 \right]^{-1} . \tag{6.7}$$

This formula obviously has no conditions for the Bose–Einstein condensation of magnons.

6.1.2
Mathematical Analogy with BEC

There are mathematical analogies (see, for example [116, 117]) for the phase transitions in magnetic systems, which formally resembles BEC of magnons. Recently some analogies (frequently called in the literature "field-induced Bose–Einstein condensation of magnons") for the "order-to-order" phase transitions have become popular in application to real magnetic systems (see, for example [118, 119]). A brief discussion of this subject is given in [120, 121].

A simple illustration of the mathematical analogy can be presented by the following example. We can rewrite (6.7) in the form

$$N = \sum_k \left[\exp\left(\frac{\hbar(\omega_k - \omega_0) - \mu_f}{k_B T} \right) - 1 \right]^{-1} , \quad \mu_f \equiv -\hbar\omega_0 . \tag{6.8}$$

Consider somewhere in the vicinity of the phase transition. The minimum of the spin-wave spectrum (usually at $k = 0$) approaches zero ($\omega_0 \to +0$). The system of magnons with the spectrum ω_k and zero chemical potential can be replaced by a fictitious "magnon system" with the gapless spectrum $\omega_k - \omega_0 \propto k^2$ and fictitious "chemical potential" $\mu_f(H)$, which depends on the external magnetic field. So far as the change of total magnon number N is very small (and therefore can be neglected), the situation is quite analogous mathematically to Bose–Einstein condensation of $k = 0$ "magnons". Physically, as we mentioned above, the magnetic excitations (magnons) with zero energy have no meaning (in comparison with the case of real particles). However, a "condensate" of fictitious $k = 0$ magnons in this picture can be considered as a finite rotation of the magnetization at the point of field-induced phase transition.

Thus we have answered the above question. There is no Bose–Einstein condensation of magnons at the thermodynamical equilibrium and the "field-induced Bose–Einstein condensation of magnons" is just a nice mathematical interpretation of phase transitions in a magnetic system, and has nothing to do with real quasi particles, magnons, the quanta of spin waves.

Generally speaking, the impossibility of Bose–Einstein condensation in systems of Bose quasi particles at thermodynamic equilibrium with a thermal bath was clear from the beginning of the concept of quasi particles. The density of quasi particles varies and reduces to zero at temperature of absolute zero. It is however possible for quasi particles to form a thermodynamic quasi equilibrium, which can lead to artificial growth and support of density of quasi particles at certain levels.

6.2
Quasi Equilibrium Magnons

As we have already demonstrated, Bose–Einstein condensation – the macroscopic quantum phenomenon in magnon gas – can occur in quasi equilibrium thermodynamic conditions which result from strong parametric pumping of magnons. In this section we shall consider the system of quasi equilibrium magnons in detail.

The interest in quasi equilibrium quasi particles has grown since the 1960s. Excitons, electron-hole bound pairs in semiconductors, became one of the first candidates for Bose–Einstein condensation at external pumping (see, for example, [122, 123]). As quasi particles, however, excitons have some problems: they have very short lifetimes ($\sim$ 1 ps) and at large densities ($\sim 10^{19}\,\mathrm{cm}^{-3}$) they can fail to be "good bosons".

There exist interesting opportunities for observation of quasi particle quasi equilibrium and the resulting Bose–Einstein condensation at ultralow temperatures in nuclear spin systems, in which the elementary nuclear spin excitations have a long lifetime. The coherent nuclear spin precession, which was discovered in 3He-B [124, 125] and was known as the homogeneously precessing domain, recently was interpreted in terms of nuclear magnon Bose–Einstein condensation [126, 127]. Another attractive prospect is the emergence of magnetic order in a nuclear spin system during adiabatic demagnetization in a solid dielectric [128]. The transition to the ordered state with tilted magnetization can be interpreted as a Bose–Einstein condensation of nuclear magnons in this system.

Magnons, as conventional elementary excitations of magnetically ordered electronic spins, have relatively long lifetimes ($\geq$ 1 μs). There was some concern that magnon systems can have high levels of nonlinear magnon–magnon interactions and therefore one can expect that the ideal gas approximation is not applicable to this type of magnon density. However, as it will be shown later, the principal nonlinearities in the system conserve the number of magnons and do not change the Bose distribution function. Magnons are still "good bosons" at high densities ($\sim 10^{22}\,\mathrm{cm}^{-3}$). The phenomenon of Bose–Einstein condensation of quasi equilibrium magnons was found theoretically at the end of the 1980s [101–103, 129, 130]. In particular, it was shown that in magnetic systems under strong microwave (either coherent or noisy) pumping there are conditions when the gas of elementary magnetic excitations can be described thermodynamically with some effective temperature and chemical potential. Bose–Einstein condensation of quasi equilibrium magnons occurs when the chemical potential approaches the bottom of the magnon spectrum. The applicability of theoretical results was discussed both for low and high temperatures. Experimental evidence in monocrystals and thin films under parametric pumping showed that there was a growth in nonequilibrium magnon population in the vicinity of the bottom of their spectrum [104–106], and this growth could not be explained in the framework of classical theory [131, 132]. The detailed experimental study became possible a decade later with the help of a new experimental technique, which allowed for the observation of the magnon population at different frequencies and wave vectors, and led to the experimental discovery of Bose–Einstein condensation of quasi equilibrium magnons [107–109, 133, 134].

6.2.1
Ideal Gas of Quasi Equilibrium Magnons

Let us consider magnons in a magnetic sample as an isolated ideal Bose gas which is in a thermodynamic quasi equilibrium with population n_k (6.1). The relation

$$N(\mu, T_{\text{eff}}) = \sum_k \left[\exp\left(\frac{\hbar\omega_k - \mu}{k_B T_{\text{eff}}}\right) - 1\right]^{-1} \tag{6.9}$$

binds together the total number of quasi equilibrium magnons N, the chemical potential μ and the effective temperature T_{eff}.

The condition $n_k \to \infty$, that defines the magnon condensation, can be written as

$$\mu_c = \min(\hbar\omega_k)\,, \quad T_{\text{eff}} > 0 \tag{6.10}$$

or, in principle,

$$\mu_c = \max(\hbar\omega_k)\,, \quad T_{\text{eff}} < 0\,. \tag{6.11}$$

Taking into account the condition of Bose–Einstein condensation $\mu = \mu_c$, from (6.9) we obtain the critical density of magnons

$$\frac{N_c}{V_s} = \frac{N(\mu_c, T_{\text{eff}})}{V_s}\,. \tag{6.12}$$

This critical number contains both the initial density of thermal magnons $N(0, T)/V_s$ and the density N_p/V_s of magnons pumped, for example, by the external microwave field. Thus, the critical density of pumped magnons is

$$\frac{N_{p,c}}{V_s} = \frac{N(\mu_c, T_{\text{eff}})}{V_s} - \frac{N(0, T)}{V_s}\,. \tag{6.13}$$

If the density of pumped magnons reached $N_{p,c}/V_s$ at given T and T_{eff}, we could expect infinite accumulation of magnons at the edge of the spectrum. In general, this principal equation for Bose–Einstein condensation of quasi equilibrium magnons can be solved numerically. For a ferromagnetic film such an estimate in [107] showed that the magnon pumping can reach the critical density $N_{p,c}/V_s$ of pumped magnons at room temperature in accordance with the experimental demonstration of Bose–Einstein condensation of quasi equilibrium magnons.

6.2.2
Example: Isotropic Spectrum

In order to simplify estimates, let us consider a ferromagnet with an isotropic magnon spectrum

$$\omega_k = \omega_0 + \nu_m k^2\,, \tag{6.14}$$

where ν_m describes the nonuniform exchange parameter.

Let us also assume that the effective temperature of quasi equilibrium magnon gas is practically equal to the thermal bath temperature $T_{eff} \simeq T$. From (6.9) we obtain the following quasi equilibrium magnon density

$$\frac{N(\mu,T)}{V_s} = \int_0^\infty \left[\exp\left(\frac{\hbar\omega_0 + \nu_m k^2 - \mu}{k_B T}\right) - 1\right]^{-1} \frac{4\pi k^2 dk}{(2\pi)^3}$$
$$= \frac{1}{8\pi^{3/2}}\left(\frac{k_B T}{\nu_m \hbar}\right)^{3/2} \mathrm{Li}_{3/2}\left[\exp\left(-\frac{\hbar\omega_0 - \mu}{k_B T}\right)\right]. \quad (6.15)$$

Taking into account a typical experimental condition $\hbar\omega_0 \ll k_B T$, one can find that

$$\mathrm{Li}_{3/2}\left[\exp\left(-\frac{\hbar\omega_0 - \mu}{k_B T}\right)\right]$$
$$\simeq \mathrm{Li}_{3/2}\left[1 - \frac{\hbar\omega_0 - \mu}{k_B T}\right]$$
$$\simeq \zeta\left(\frac{3}{2}\right) - 2\sqrt{\pi\frac{\hbar\omega_0 - \mu}{k_B T}}. \quad (6.16)$$

As a result we can rewrite (6.15) in the form

$$\frac{N(\mu,T)}{V_s} = \frac{1}{8\pi^{3/2}}\left(\frac{k_B T}{\nu_m \hbar}\right)^{3/2}\left[\zeta\left(\frac{3}{2}\right) - 2\sqrt{\pi\frac{\hbar\omega_0 - \mu}{k_B T}}\right]. \quad (6.17)$$

From this formula we can get the initial density of thermal magnons

$$\frac{N(0,T)}{V_s} = \frac{1}{8\pi^{3/2}}\left(\frac{k_B T}{\nu_m \hbar}\right)^{3/2}\left[\zeta\left(\frac{3}{2}\right) - 2\sqrt{\frac{\pi\hbar\omega_0}{k_B T}}\right] \quad (6.18)$$

and the critical magnon density at the condition of condensation $\mu_c = \hbar\omega_0$:

$$\frac{N_c}{V} = \frac{1}{8}\zeta\left(\frac{3}{2}\right)\left(\frac{k_B T}{\pi\nu_m \hbar}\right)^{3/2}. \quad (6.19)$$

Note that the last formula does not contain the energy gap $\hbar\omega_0$. It can also be rewritten in the form

$$T = \frac{4\pi}{\zeta^{2/3}(3/2)}\frac{\nu_m \hbar}{k_B}\left(\frac{N_c}{V_s}\right)^{2/3}, \quad (6.20)$$

which, with the accuracy of notations, gives the relation between the temperature and the critical density of Bose–Einstein condensation. This formula looks similar to the formula for the relation between T_{BEC} and boson density in the equilibrium case (6.6). However, (6.20) can be used in this sense only in the case when the density of magnons mostly results from pumping: $N_c/V_s \simeq N_p/V_s \gg N(0,T)/V_s$. In the most interesting case of high temperatures, when the initial density of thermal magnons is high, we have $N_c/V_s \simeq N(0,T)/V_s \gg N_p/V_s$. Therefore, formula (6.20) in this case does not contain any information about pumped magnons and becomes useless for a characterization of Bose–Einstein condensation of quasi equilibrium magnons.

From (6.18) and (6.19) we can calculate the density of pumped magnons N_p/V_s and the condition of quasi equilibrium Bose–Einstein condensation can be written in the form:

$$\frac{N_{p,c}}{V_s} = \frac{k_B T}{4\pi\hbar}\frac{\omega_0^{1/2}}{\nu_m^{3/2}} . \tag{6.21}$$

Let us make estimates for YIG, in which $\nu_m = 0.092\,\mathrm{cm}^{-2}\mathrm{s}^{-1}$. For magnon gas with the frequency gap of $\omega_0/2\pi = 2.4\,\mathrm{GHz}$ at room temperature $T = 300\,\mathrm{K}$ we obtain $N_{p.c}/V_s \sim 1.4\times10^{19}\,\mathrm{cm}^{-3}$, $N_c/V_s \sim 0.52\times10^{21}\,\mathrm{cm}^{-3}$ and $N_{p.c}/N_c \sim 0.027$.

From (6.20) and (6.21) we can also obtain

$$\frac{N_{p,c}}{N_c} = 2\zeta^{-1}\left(\frac{3}{2}\right)\sqrt{\frac{\pi\hbar\omega_0}{k_B T}} , \tag{6.22}$$

which says that the relative part of pumped magnons to the total number of magnons decreases with increasing temperature.

Problem 6.1. Estimate N_c/V_s and $N_{p.c}/V_s$ for the magnon isotropic spectrum $\omega_k = \sqrt{\omega_0^2 + (\nu k)^2}$ of an antiferromagnet with "easy plane" anisotropy.

6.2.3 Kinetic Equations

Now we describe the evolution of a magnon system under incoherent pumping. Let us consider a kinetic equation for boson occupation numbers

$$\frac{d}{dt} n_{\boldsymbol{k}} = I^{(4)}\{n_{\boldsymbol{k}}\} + f_{\boldsymbol{k}} - r_{\boldsymbol{k}} , \tag{6.23}$$

where $f_{\boldsymbol{k}}$ describes the probability to excite a magnon by the external electromagnetic field and $r_{\boldsymbol{k}}$ is the relaxation term responsible for magnon interaction with a thermal bath (phonons, and so on). The four-magnon collision integral is defined by

$$\begin{aligned} I^{(4)}\{n_{\boldsymbol{k}}\} = \frac{2\pi}{\hbar^2} &\sum_{\boldsymbol{k}_1,\boldsymbol{k}_2,\boldsymbol{k}_3} |\Phi(\boldsymbol{k},\boldsymbol{k}_1;\boldsymbol{k}_2,\boldsymbol{k}_3)|^2 \\ &\times [(n_{\boldsymbol{k}}+1)(n_{\boldsymbol{k}_1}+1)n_{\boldsymbol{k}_2}n_{\boldsymbol{k}_3} - n_{\boldsymbol{k}}n_{\boldsymbol{k}_1}(n_{\boldsymbol{k}_2}+1)(n_{\boldsymbol{k}_3}+1)] \\ &\times \Delta(\boldsymbol{k}+\boldsymbol{k}_1-\boldsymbol{k}_2-\boldsymbol{k}_3)\delta(\omega_{\boldsymbol{k}}+\omega_{\boldsymbol{k}_1}-\omega_{\boldsymbol{k}_2}-\omega_{\boldsymbol{k}_3}) , \end{aligned} \tag{6.24}$$

where $\Phi(\boldsymbol{k}, \boldsymbol{k}_1; \boldsymbol{k}_2, \boldsymbol{k}_3)$ is the magnon–magnon scattering amplitude, $n_{\boldsymbol{k}}$ is the magnon population, and $\omega_{\boldsymbol{k}}$ is the magnon frequency.

These three terms are characterized by different time scale parameters. Magnon–magnon exchange scattering makes a substantial contribution to the kinetics of the entire magnon system. Below we assume that the typical time τ_4 of magnon–magnon scattering is much less than the characteristic time τ_s of magnon excitation (i.e., energy supply) and characteristic time τ_r of relaxation of number of magnons. In other words, it is assumed that magnon–magnon collision is the principal process in the system, while interaction of magnons with the pump field and the heat bath are considerably less probable to the extent of the smallness of the parameters $\tau_4/\tau_s \ll 0$ and $\tau_4/\tau_r \ll 0$, respectively. This means that the kinetic (6.23) in the stationary case in the zeroth approximation with respect to these small parameters can be written as

$$0 = I^{(4)}\{n_{\boldsymbol{k}}\} . \tag{6.25}$$

The formal solution of (6.25) is the Bose quasi equilibrium distribution function

$$n_k = \left[\exp\left(\frac{\hbar\omega_{\boldsymbol{k}} - \mu}{k_B T_{\text{eff}}}\right) - 1\right]^{-1} , \tag{6.26}$$

with the effective chemical potential μ and the effective temperature T_{eff}. These two parameters define the density of quasi equilibrium magnons and their energy density:

$$\begin{aligned} \frac{N}{V_s} &= \int n_{\boldsymbol{k}} \frac{d^3k}{(2\pi)^3} , \\ \frac{\mathcal{E}}{V_s} &= \int \hbar\omega_{\boldsymbol{k}} n_{\boldsymbol{k}} \frac{d^3k}{(2\pi)^3} . \end{aligned} \tag{6.27}$$

On the other hand, these equations can be taken as definitions for $\mu(N/V_s, \mathcal{E}/V_s)$ and $T_{\text{eff}}(N/V_s, \mathcal{E}/V)$, or $\mu(N/V_s, T_{\text{eff}})$ and $\mathcal{E}(N/V_s, T_{\text{eff}})$.

To determine explicitly the evolution of μ and T_{eff}, one can write the balance equations for the total number of magnons and the energy of magnon system that can be obtained from (6.23) by summing over $\boldsymbol{k}$:

$$\frac{d}{dt}N = \mathcal{F}_N - \mathcal{R}_N , \tag{6.28}$$

$$\frac{d}{dt}\mathcal{E} = \mathcal{F}_{\mathcal{E}} - \mathcal{R}_{\mathcal{E}} . \tag{6.29}$$

Here the following notations have been introduced:

$$\mathcal{F}_N \equiv V_s \int f_k \frac{d^3k}{(2\pi)^3} , \quad \mathcal{R}_N \equiv V_s \int r_k \frac{d^3k}{(2\pi)^3} , \tag{6.30}$$

$$\mathcal{F}_{\mathcal{E}} \equiv V_s \int \varepsilon_k f_k \frac{d^3k}{(2\pi)^3} , \quad \mathcal{R}_{\mathcal{E}} \equiv V_s \int \varepsilon_k r_{\boldsymbol{k}} \frac{d^3k}{(2\pi)^3} . \tag{6.31}$$

A complete rigorous analysis of (6.28) and (6.29) is complicated and depends on scattering amplitudes and relaxation terms $\mathcal{R}_N$ and $\mathcal{R}_{\mathcal{E}}$. For simplicity and to preserve the general character of our analysis, we confine ourselves to the case of τ-approximation:

$$\mathcal{R}_N = \frac{N(\mu, T_{\text{eff}}) - N(0, T_{\text{eff}})}{\tau_N(\mu, T_{\text{eff}})}, \tag{6.32}$$

$$\mathcal{R}_{\mathcal{E}} = \frac{\mathcal{E}(\mu, T_{\text{eff}}) - \mathcal{E}(\tilde{\mu}, T)}{\tau_{\mathcal{E}}(\mu, T)}, \tag{6.33}$$

where the parameter τ_N^{-1} is the rate of change of the total number of magnons, and $\tau_{\mathcal{E}}^{-1}$ is the rate of change of the Bose-system energy. The deviations of N and $\mathcal{E}$ are known from the quasi equilibrium state of a strongly excited Bose system, and the formal parameter $\tilde{\mu}$ is determined from the condition

$$N(\tilde{\mu}, T) = N(\mu, T_{\text{eff}}) . \tag{6.34}$$

The origin of this equality is that the magnon temperature is established in a time τ_4 much shorter than τ_N.

It is easy to verify that in the absence of an external pump (6.28)–(6.34) lead to the solution $\mu = 0$, $T = T_B$, that is, to the exact solution of the stationary state kinetic equation

$$0 = I^{(4)}\{n_k\} - r_k \tag{6.35}$$

without energy supply.

6.2.3.1 The Case of $T_{\text{eff}} = T$

In the simplest case when $\tau_{\mathcal{E}} \ll \tau_N$ the change of the magnon system temperature can be neglected. Then the analysis is reduced to (6.28) for the chemical potential.

As we have discussed earlier, the simplest method to excite magnons in magnets is via decay of a microwave field quantum into a pair of magnons of half the frequency and with equal but opposite wave vectors. In this case the probability to excite magnons is defined by

$$\begin{aligned} f_k(\mu, T) &= \frac{2\pi}{\hbar} \int \varphi(\omega) |h V_k|^2 [(n_k + 1)(n_{-k} + 1) - n_k n_{-k}] \delta(\omega - 2\omega_k) d\omega \\ &= \frac{2\pi}{\hbar} \varphi(2\omega_k) |h V_k|^2 [2 n_k(\mu, T) + 1] , \end{aligned} \tag{6.36}$$

where h is the microwave field amplitude and $\varphi(\omega)$ is the noise pump line shape normalized to unity.

We can write the balance equation

$$\mathcal{F}_N = \mathcal{R}_N \tag{6.37}$$

in the explicit form

$$2\pi \int \varphi(2\omega_k)|h V_k|^2[2n_k(\mu, T) + 1]\frac{d^3k}{(2\pi)^3}$$
$$= \tau_N^{-1}(\mu, T) \int \left[n_k(\mu, T) - n_k(0, T)\right] \frac{d^3k}{(2\pi)^3} \,. \tag{6.38}$$

Consider a ferromagnet in which the coefficient of magnon coupling with the external field is (e.g., [73]):

$$V_k = \frac{2\pi\gamma^2 M_s}{\omega_k} \sin^2\theta_k \exp(i2\phi_k) \tag{6.39}$$

and the magnon spectrum is defined by

$$\omega_k \simeq \omega_0 + \nu_m k^2 + 2\pi\gamma M_s \sin^2\theta_k \,. \tag{6.40}$$

Here M_s is the magnetization, θ_k and ϕ_k are the polar and azimuthal angles of the vector $\boldsymbol{k}$ in a spherical coordinate system whose axis is oriented along $\boldsymbol{M}_s$.

To be specific, we consider the incoherent pump having an equal alternating magnetic field amplitude h in the frequency range from Ω_{ext} to $\Omega_{ext} + \Delta\Omega$ ($\Omega_{ext} > 2\omega_0$):

$$\varphi(\Omega) = \begin{cases} \Delta\Omega^{-1}, & \Omega \in [\Omega_{ext}, \Omega_{ext} + \Delta\Omega]\,, \\ 0\,, & \Omega \notin [\Omega_{ext}, \Omega_{ext} + \Delta\Omega]\,. \end{cases} \tag{6.41}$$

We then calculate

$$N(\mu, T) = \frac{V_s}{(2\pi)^3} I_{1f}(T, \mu)\,, \tag{6.42}$$

$$I_{1f}(T, \mu) = \int_0^\infty dk \int_{-1}^{1} d\cos\theta \; k^2$$
$$\times \left\{ \left[\exp\left(\frac{\hbar\omega_k - \mu}{k_B T}\right) - 1\right]^{-1} - \left[\exp\left(\frac{\hbar\omega_k}{k_B T}\right) - 1\right]^{-1} \right\}$$

and

$$\mathcal{F}_N(\mu, T) \simeq \frac{2\pi V_s}{\Delta\Omega} \left[\gamma^2 h M_s\right]^2 I_{2f}(T, \Omega_{ext}, \Delta\Omega)\,, \tag{6.43}$$

$$I_{2f}(T, \Omega, \Delta\Omega) = \int_{\Omega/2 \le \omega_k \le \Omega + \Delta\Omega/2} dk \int d\cos\theta \, k^2 \left(\frac{k_B T}{\hbar\omega_k}\right)^2$$
$$\times \left\{ 2\left[\exp\left(\frac{\hbar\omega_k}{k_B T}\right) - 1\right]^{-1} + 1 \right\} .$$

As a result, from (6.38) one obtains

$$(\gamma h)^2 = \frac{\Delta\Omega\, \tau_N^{-1}}{(2\pi)^3 (\gamma M_s)^2} \left(\frac{k_B T}{\hbar}\right)^2 \frac{I_{1f}(T, \mu) - I_{1f}(T, 0)}{I_{2f}(T, \Omega_{ext}, \Delta\Omega)} \,. \tag{6.44}$$

6.2.4
Magnon System with Bose Condensate

Now let us consider the state of a magnon system in which the effective chemical potential has already reached the bottom of the magnon band. We write the magnon system Hamiltonian in the form

$$\mathcal{H} = \sum_{k}(\hbar\omega_k - \mu) b_k^\dagger b_k + \frac{\hbar}{2}\sum_{1,2,3,4} \Phi(1,2;3,4) b_1^\dagger b_2^\dagger b_3 b_4 \Delta(\boldsymbol{k}_1 + \boldsymbol{k}_2 - \boldsymbol{k}_3 - \boldsymbol{k}_4) , \tag{6.45}$$

where the first term describes an ideal Bose gas and the second one is responsible for magnon–magnon scattering processes. The terms describing explicitly the noise pump and the energy loss of the system have been omitted. We confine ourselves to an approximation in which the role of the pumping reduces only to changing the number of Bose quasi particles, and can be taken into account via the function of the total magnon number

$$N = \sum_{k} b_k^\dagger b_k . \tag{6.46}$$

This number is defined by the balance between the pumping and the relaxation of magnons. To take into account the constancy of the quasi particle number (for a given pump level), we have introduced into (6.45) the chemical potential μ of the magnons.

Following the theory of weakly ideal Bose gas, developed for gap-free spectra, we single out the classical condensate amplitudes with $\boldsymbol{k} = \boldsymbol{0}$:

$$b_0^\dagger = b_0 = \sqrt{N_0} , \quad N_0 = N - N' . \tag{6.47}$$

Assuming that $N'/N \ll 1$, we obtain then from (6.45) the Hamiltonian with the accuracy up to quadratic terms

$$\mathcal{H} = (\hbar\omega_0 - \mu) N_0 + \frac{1}{2}\hbar \mathcal{T}_{00} N_0^2 + \sum_{k \neq 0}\left[\mathcal{A}_k b_k^\dagger b_k + \frac{\mathcal{B}_k}{2}\left(b_k b_{-k} + b_k^\dagger b_{-k}^\dagger\right)\right] , \tag{6.48}$$

where

$$\mathcal{A}_k = \hbar\omega_k - \mu + 2\hbar \mathcal{T}_{0k} N_0 ,$$
$$\mathcal{B}_k = \mathcal{S}_{0k} N_0 ,$$

and

$$\mathcal{T}_{0k} \equiv \Phi(\boldsymbol{0}, \boldsymbol{k}; \boldsymbol{0}, \boldsymbol{k}) , \quad \mathcal{S}_{0k} \equiv \Phi(\boldsymbol{0}, \boldsymbol{0}; \boldsymbol{k}. - \boldsymbol{k}) .$$

Diagonalizing the quadratic form (6.48) by the linear transformation

$$b_k = u_k c_k + v_k c^\dagger_{-k} \,, \qquad b^\dagger_k = u_k c^\dagger_k + v_k c_{-k} \,, \tag{6.49}$$

we readily find that

$$\mathcal{H} = \mathcal{U}_0 + \sum_k \hbar \tilde{\omega}_k c^\dagger_k c_k \,, \tag{6.50}$$

where

$$\mathcal{U}_0 = (\hbar\omega_0 - \mu) N_0 + \frac{1}{2}\hbar \mathcal{T}_{00} N_0^2 + \frac{1}{2}\sum_{k\neq 0} (\hbar\tilde{\omega}_k - \mathcal{A}_k)$$

is the energy of the ground state, in which the chemical potential μ, the number of magnons N_0, and the spectrum of the new quasi particles $\tilde{\omega}_k$ are defined by the following equations:

$$\mu = \hbar\omega_0 + \hbar \mathcal{T}_{00} N_0 + \sum_{k\neq 0} \left\{ \mathcal{T}_{0k} \left[\frac{\mathcal{A}_k}{\tilde{\omega}_k} \coth\left(\frac{\hbar\tilde{\omega}_k}{2k_\mathrm{B}T}\right) - 1 \right] - \frac{\mathcal{T}_{0k} N_0^2}{2\hbar\tilde{\omega}_k} \coth\left(\frac{\hbar\tilde{\omega}_k}{2k_\mathrm{B}T}\right) \right\} \,, \tag{6.51}$$

$$N = N_0 + \frac{1}{2}\sum_{k\neq 0} \left[\frac{\mathcal{A}_k}{\hbar\tilde{\omega}_k} \coth\left(\frac{\hbar\tilde{\omega}_k}{2k_\mathrm{B}T}\right) - 1 \right] \,, \tag{6.52}$$

and

$$\hbar\tilde{\omega}_k = \left[(\hbar\omega_k - \mu + 2\mathcal{T}_{0k} N_0)^2 - (\mathcal{S}_{0k} N_0)^2 \right]^{1/2} \,. \tag{6.53}$$

In the zeroth approximation with respect to $N'/N \ll 1$, the new quasi particle spectrum can be written in the form

$$\tilde{\omega}_k = \left[N^2 \left(\mathcal{T}_{0k}^2 - \mathcal{S}_{0k}^2\right) + 2N\mathcal{T}_{0k} (\omega_k - \omega_0) \right]^{1/2} \,. \tag{6.54}$$

Thus, for all $\boldsymbol{k}$ the condition of stability is

$$N \left(\mathcal{T}_{0k}^2 - \mathcal{S}_{0k}^2\right) + 2\mathcal{T}_{0k} (\omega_k - \omega_0) \geq 0 \,. \tag{6.55}$$

6.2.5 Magnetodipole Emission of Condensate

Let us consider the coherent emission of photons by the condensate with $\boldsymbol{k} = 0$. The magnetodipole emission intensity is given by the formula [135]:

$$\mathcal{I} = \frac{2}{3c^3} \left(\frac{d^2 \boldsymbol{M} V_\mathrm{s}}{dt^2} \right)^2 \tag{6.56}$$

where $\boldsymbol{M} V_\mathrm{s}$ is the magnetic moment of the sample and c is the speed of light.

The nonlinear character of the magnetic moment motion produces in the emission spectrum, besides the fundamental frequency $\omega = \omega_0$, the harmonics 2ω, 3ω, and so on. As a result

$$\mathcal{I} = \mathcal{I}_\omega + \mathcal{I}_{2\omega} + \mathcal{I}_{3\omega} + \ldots \tag{6.57}$$

where $\mathcal{I}_{n\omega}$ is the nth harmonic radiation intensity.

Assuming that the photon emission is the principal process of relaxation, we can estimate the upper limit of the emission intensity from the balance equation of (6.37)

$$\mathcal{F}_N \simeq \sum_{n=1} \frac{\mathcal{I}_{n\omega}}{n\hbar\omega} = \mathcal{R}_N \tag{6.58}$$

in which $\mathcal{F}_N$ is determined by (6.36) and (6.30) with $n_k \simeq 0$.

For a cubic ferromagnet one obtains

$$\begin{aligned} \mathcal{I}_\omega &= \alpha_m \omega^4 S \mathcal{N} N_0 \,, \\ \mathcal{I}_{2\omega} &= \alpha_m \nu^2 (2\omega)^4 \, N_0^2 \,, \\ \mathcal{I}_{3\omega} &= \alpha_m \nu^2 (3\omega)^4 \frac{N_0^3}{16 S \mathcal{N}} \,, \end{aligned} \tag{6.59}$$

where

$$\alpha_m = \frac{2^4 \mu_\mathrm{B}^2}{3c^3} \,,$$

S is the electronic spin, $\mathcal{N}$ is the number of elementary cells in the sample,

$$\begin{aligned} \nu^2 &= \frac{A_0 - \omega_0}{2\omega_0} \,, \\ A_0 &= 2\pi\gamma M_\mathrm{s} \left(N_{xx} + N_{yy} - 2N_{zz}\right) \,, \\ \omega_0 &= \gamma \left[(H_\mathrm{i} + 2\pi M_\mathrm{s} N_{xx}) \left(H_\mathrm{i} + 4\pi M_\mathrm{s} N_{yy}\right)\right]^{1/2} \,, \end{aligned}$$

N_{xx}, N_{yy}, N_{zz} are the demagnetizing factors of the sample, and $H_\mathrm{i} = H - 4\pi M_\mathrm{s} N_{zz}$ is the internal field. We see that by varying the sample geometry (and therefore changing the demagnetizing factors) it is possible to strongly change the emission intensity at a given frequency.

The relations (6.59), however, may be violated if the sample is placed in a resonator cavity tuned to one of the radiation frequencies. It is particularly interesting that the role of the resonator can be played by the intrinsic magnetostatic mode of the sample.

6.3
Fröhlich Coherence

Coherence is a result of of phase relationships, which are readily destroyed by perturbations. For this reason superconducting and superfluid states of matter, demonstrating macroscopic coherence, exist only in the relative absence of thermal fluctuations. Such states exhibit only the simplest kind of phase relationships, which are coupled to the environment.

Complex dynamical systems have subtle internal phase relationships, and in some cases the nature of the dynamics protects these relationships through feedback, amplification, and so on, especially in the presence of a supply of energy.

Herbert Fröhlich, one of the great pioneers in physics, proposed a model [136, 137] of a system of coupled molecular oscillators in a heat bath, supplied with energy at a constant rate. When this rate exceeds a certain threshold, a condensation of the whole system of oscillators takes place into one giant dipole mode, just like in Bose–Einstein condensation. In this section we shall consider this famous model.

Any system containing charged particles is capable of long-wave electrical vibrations. Such a system need not be homogeneous. One characteristic of these vibrations is the existence of a lowest frequency $\omega_1 \neq 0$. The system is assumed to possess N_{osc} modes: $\omega_1 < \omega_2 < \cdots < \omega_{N_{\text{osc}}}$. They do not to interact with each other but are strongly coupled to the heat bath at temperature T.

Let n_k be the number of quanta $\hbar\omega_k$ in the mode ω_k. We consider the following kinetic equation:

$$\frac{dn_k}{dt} = f_k - \phi\left\{n_k\left[\exp\left(\frac{\hbar\omega_k}{k_B T}\right) - 1\right] - 1\right\} - \chi\sum_j\left[n_k(1+n_j)\exp\left(\frac{\hbar\omega_k}{k_B T}\right) - n_j(1+n_k)\exp\left(\frac{\hbar\omega_j}{k_B T}\right)\right] . \tag{6.60}$$

Here f_k is the rate of energy supply to the k-mode. Coefficient ϕ describes the rate of the first-order relaxation processes in which single quanta $\hbar\omega_k$ may be emitted to or absorbed from the heat bath. The second-order processes are characterized by the rate χ and give nonlinearity in the system. Note that the second-order processes lead to interchange of quanta $\hbar\omega_k$ without alteration of their total number

$$N = \sum_k n_k .$$

The general forms of the first-order and the second-order processes are dictated by the requirement that in the absence of energy supply, $f_k = 0$, thermal equilibrium requires a Planck distribution for n_k.

We wish to solve (6.60) for the stationary case $dn_k/dt = 0$. Then

$$\begin{aligned} 0 &= f_k - \phi\left[n_k\exp\left(\frac{\hbar\omega_k}{k_B T}\right) - (n_k+1)\right] \\ &\quad - \chi\sum_j\left[n_k(1+n_j)\exp\left(\frac{\hbar\omega_k}{k_B T}\right) - n_j(1+n_k)\exp\left(\frac{\hbar\omega_j}{k_B T}\right)\right] \\ &= f_k + \mathcal{B} - n_k\left[\mathcal{A}\exp\left(\frac{\hbar\omega_k}{k_B T}\right) - \mathcal{B}\right] , \end{aligned} \tag{6.61}$$

where

$$\mathcal{A} = \phi + \chi \sum_j (1 + n_j) ,$$

$$\mathcal{B} = \phi + \chi \sum_j n_j \exp\left(\frac{\hbar\omega_j}{k_B T}\right) . \tag{6.62}$$

From (6.61) it formally follows that

$$n_k = \frac{f_k + \mathcal{B}}{\mathcal{A} \exp\left(\frac{\hbar\omega_k}{k_B T}\right) - \mathcal{B}} . \tag{6.63}$$

Summation (6.61) over k with the definition,

$$F \equiv \sum_k f_k$$

yields

$$F = \phi \sum_k \left[n_k \exp\left(\frac{\hbar\omega_k}{k_B T}\right) - (n_k + 1) \right]$$

$$= \phi \left[\sum_k n_k \exp\left(\frac{\hbar\omega_k}{k_B T}\right) - N - N_{\text{osc}} \right] , \tag{6.64}$$

where $N_{\text{osc}} = \sum_k 1$.

Now we define μ by

$$\exp\left(-\frac{\mu}{k_B T}\right) = \frac{\phi + \chi \sum_j (1 + n_j)}{\phi + \chi \sum_j n_j \exp\left(\frac{\hbar\omega_j}{k_B T}\right)} = \frac{\mathcal{A}}{\mathcal{B}} , \tag{6.65}$$

so that

$$1 - \exp\left(-\frac{\mu}{k_B T}\right) = \frac{\chi \sum_k \left[n_k \exp(\hbar\omega_k / k_B T) - (n_k + 1) \right]}{\phi + \chi \sum_j n_j \exp(\hbar\omega_j / k_B T)}$$

$$= \frac{\chi (F/\phi)}{\phi + \chi \sum_j n_j \exp(\hbar\omega_j / k_B T)}$$

$$= \frac{\chi F}{\phi \mathcal{B}} \geq 0 . \tag{6.66}$$

We can see that μ increases with increasing energy supply F. Solving (6.65) and (6.66), and substituting $\mathcal{A}$, $\mathcal{B}$ into (6.63), one obtains

$$n_k = \left\{ 1 + \frac{\phi f_k}{\chi F} \left[1 - \exp\left(-\frac{\mu}{k_B T}\right) \right] \right\} \left[\exp\left(\frac{\hbar\omega_k - \mu}{k_B T}\right) - 1 \right]^{-1} . \tag{6.67}$$

It follows then that a single mode ω_1 gets very strongly (coherently) excited when the energy supply reaches a critical value F_c which corresponds to

$$\mu(F_c) = \hbar\omega_1 . \tag{6.68}$$

The supplied energy is thus not completely thermalized but stored in a highly ordered fashion. This order expresses itself in long-range phase correlations; the phenomenon has considerable similarity with the low-temperature condensation of a Bose gas.

Fröhlich's model gives a clue to coherence in a living object (proteins, cells, viruses, and so on) but has been considered for some time only as a hypothesis since it is difficult to believe that a Bose–Einstein condensation, a phenomenon observed in liquid helium at very low temperatures, can occur at high temperatures. Now the situation has changed due to the discovery of Bose–Einstein condensation of quasi equilibrium magnons at room temperature. Theoretically, this phenomenon in some sense resembles Fröhlich's model. We have a real physical system which exhibits Bose–Einstein condensation at room temperature, and therefore, one can seriously expect that coherent phenomena could perhaps occur in biological systems as well.

6.4 Discussion

a) Does Bose–Einstein condensation of magnons occur at the equilibrium conditions?
b) What is a chemical potential of a magnon?
c) What is the Bose–Einstein condensate stability criterion?
d) Does a uniform rotation of magnetization resemble a magnon condensate?

7
Magnons in an Ultrathin Film

Magnons as elementary collective excitations of ordered spins exist in thin and ultrathin magnetic films and confined structures [138–141]. Studies of large magnetization motions, including reversal, in ultrathin ferromagnetic films under an applied external magnetic field are of great importance in application to magnetic devices, in particular, in magnetic recording physics. Large-angle magnetization motion can excite magnon instabilities, which increase substantially the magnetization reversal rate [142]. The total film magnetization $|\boldsymbol{M}|$ decreases in this case, which was observed experimentally [143].

The problem of nonlinear spin-wave excitation during reversal and large magnetization motions has been explored by numerical simulations in nanograins [144–146] and thin films [147–149]. All these simulations have been performed using conventional local micromagnetic modeling, which includes: (i) the analysis of intra- and intercell interactions, (ii) analysis of phenomenological dynamic equations, and (iii) computer simulations. There are two principal problems with this technique: (i) the physical problem of the introduction of local phenomenological damping (and corresponding magnetic noise) and (ii) the computing problem in the case of a large number of cells.

Both problems of local micromagnetic modeling can be avoided by developing $\boldsymbol{k}$-space micromagnetic modeling. In this case the theory includes: (i) an analysis of magnon spectra and interactions in an ultrathin film, (ii) the calculation of the effective scattering processes (most of the accumulated energy is to be transformed to nonlinear spin waves), and (iii) the analysis of self-consistent dynamic equations with microscopic damping (and noise, if necessary). The analytical technique to study nonlinear spin-wave dynamics has been developed in the theory of parametric magnon excitation (mainly in the bulk, see, for example, [4, 71–73, 77]). In application to magnetic thin films $\boldsymbol{k}$-space modeling has been considered in [150–153].

Here, as an example, we give an explicit analytic formulation to describe magnetization reversal (switching) for up to 90° deviation from equilibrium in ultrathin films in terms of magnon-pair excitations.

Nonequilibrium Magnons, First Edition. Vladimir L. Safonov.

7.1 Model

Let us consider an ultrathin ferromagnetic film ($\tau \times L_y \times L_z$) with the magnetization: $\boldsymbol{M}(\boldsymbol{r}) = M_s \boldsymbol{m}(\boldsymbol{r})$, $\boldsymbol{r} = (y, z)$. The variation of the unit vector $\boldsymbol{m}$ within the film thickness $(-\tau/2 \le x \le \tau/2)$ will be neglected. Locally one has:

$$|\boldsymbol{m}(\boldsymbol{r})| = \sqrt{m_x^2 + m_y^2 + m_z^2} = 1 \,.$$

In order to introduce collective magnetization motions, we assume that the film is periodic along both y and z directions with periods L_y and L_z, respectively. The Fourier series representation can be written as

$$\boldsymbol{m}(\boldsymbol{r}) = \sum_{\boldsymbol{k}} \boldsymbol{m}_{\boldsymbol{k}} \exp(i\boldsymbol{k}\cdot\boldsymbol{r}) \,,$$

$$\boldsymbol{m}_{\boldsymbol{k}} = \frac{1}{L_y L_z} \int_0^{L_y} dy \int_0^{L_z} dz\, \boldsymbol{m}(\boldsymbol{r}) \exp(-i\boldsymbol{k}\cdot\boldsymbol{r}) \,, \tag{7.1}$$

where

$$k_y = \frac{2\pi n_y}{L_y} \,, \quad k_z = \frac{2\pi n_z}{L_z} \,, \quad -\infty < n_y, n_z < \infty$$

are the wave vector components in the plane.

The equilibrium is supposed to be a uniformly magnetized state, in which the magnetization is oriented in the (y, z)-plane along an equilibrium axis z_0. The transformation from the (x, y, z) coordinates to equilibrium coordinates (Figure 7.1) (x_0, y_0, z_0) is defined by

$$\begin{pmatrix} y \\ z \end{pmatrix} = \begin{pmatrix} \cos\theta_0 & \sin\theta_0 \\ -\sin\theta_0 & \cos\theta_0 \end{pmatrix} \begin{pmatrix} y_0 \\ z_0 \end{pmatrix} \,. \tag{7.2}$$

Here θ_0 determines a rotation in the film plane, $x = x_0$. Analogous transformation should be used for $(m_y, m_z) \to (m_{y0}, m_{z0})$ and wave vector components $(k_y, k_z) \to (k_{y0}, k_{z0})$. Note that both the absolute value of the wave vector $k = |\boldsymbol{k}|$ and the scalar product $\boldsymbol{k}\cdot\boldsymbol{r}$ are invariant with respect to the choice of system of coordinates.

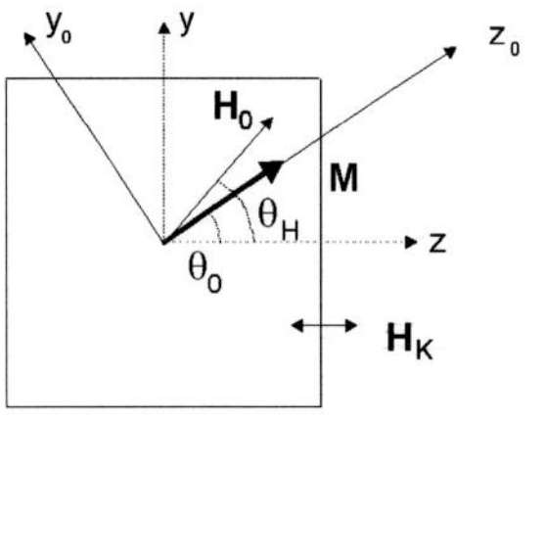

Figure 7.1 Equilibrium coordinates in the film plane.

7.1.1
Magnetic Energy

The magnetic energy of the film contains the exchange energy, energy of anisotropy, Zeeman energy and demagnetization energy:

$$\mathcal{E} = \mathcal{E}_{\text{exch}} + \mathcal{E}_{\text{anis}} + \mathcal{E}_Z + \mathcal{E}_{\text{dmag}} \,. \tag{7.3}$$

The exchange energy $-A(\nabla \cdot \boldsymbol{m})^2$ can be represented as

$$\mathcal{E}_{\text{exch}} = VA \sum_{\boldsymbol{k}} k^2 \left(m_{y0,\boldsymbol{k}} m_{y0,-\boldsymbol{k}} + m_{z0,\boldsymbol{k}} m_{z0,-\boldsymbol{k}} \right) , \tag{7.4}$$

where A is the exchange constant and $V = \tau L_y L_z$ is the film volume. To obtain (7.4) we have used the following formula:

$$\frac{1}{L_y L_z} \int_0^{L_y} dy \int_0^{L_z} dz \exp[i(\boldsymbol{k} + \boldsymbol{k}_1) \cdot \boldsymbol{r}] = \Delta(\boldsymbol{k} + \boldsymbol{k}_1) , \tag{7.5}$$

where $\Delta(\cdot)$ is the Kronecker delta function.

The quadratic uniaxial anisotropy energy (z is an easy axis, see, Figure 7.1) in $\boldsymbol{k}$-space is:

$$\begin{aligned} \mathcal{E}_{\text{anis}} = -V K_1 \sum_{\boldsymbol{k}} & \left(-m_{y0,\boldsymbol{k}} \sin\theta_0 + m_{z0,\boldsymbol{k}} \cos\theta_0 \right) \\ & \times \left(-m_{y0,-\boldsymbol{k}} \sin\theta_0 + m_{z0,-\boldsymbol{k}} \cos\theta_0 \right) . \end{aligned} \tag{7.6}$$

The Zeeman energy in the external magnetic field $\boldsymbol{H}_0 = (0, H_0 \sin\theta_H, H_0 \cos\theta_H)$ is:

$$\mathcal{E}_Z = -V M_s H_0 [m_{y0,0} \sin(\theta_H - \theta_0) + m_{z0,0} \cos(\theta_H - \theta_0)] \,. \tag{7.7}$$

The demagnetization energy (see, [156]) is defined by:

$$\begin{aligned} \mathcal{E}_{\text{dmag}} = 2\pi M_s^2 V \sum_{\boldsymbol{k}} & \Bigg\{ G(k\tau) m_{x0,\boldsymbol{k}} m_{x0,-\boldsymbol{k}} \\ & + [1 - G(k\tau)] \Bigg[\left(\frac{k_{y0}}{k} \right)^2 m_{y0,\boldsymbol{k}} m_{y0,-\boldsymbol{k}} \\ & + \left(\frac{k_{z0}}{k} \right)^2 m_{z0,\boldsymbol{k}} m_{z0,-\boldsymbol{k}} + 2 \frac{k_{y0} k_{z0}}{k^2} m_{y0,\boldsymbol{k}} m_{z0,-\boldsymbol{k}} \Bigg] \Bigg\} , \end{aligned} \tag{7.8}$$

where

$$G(x) = \frac{1 - \exp(-x)}{x} \,.$$

7.2 Magnons

We shall utilize a classical form of the spin representation in terms of Bose operators introduced in [154, 155] and convenient for two-dimensional systems. It should be noted that the Holstein–Primakoff transformation in the two-dimensional case is a less accurate spin expansion.

For the unit magnetization vector $\boldsymbol{m}$ this representation in terms of complex variables a and a^* can be written as:

$$m_{x_0} = i\frac{a - a^*}{\sqrt{2}}\,, \tag{7.9}$$

$$m_{y_0} = \sqrt{1 - m_{x_0}^2}\sin\left(\frac{a + a^*}{\sqrt{2}}\right)\,, \tag{7.10}$$

$$m_{z_0} = \sqrt{1 - m_{x_0}^2}\cos\left(\frac{a + a^*}{\sqrt{2}}\right)\,. \tag{7.11}$$

An expansion of (7.9)–(7.11) up to the fourth order gives accuracy $\sim 6\%$ for about 90° deviation from equilibrium:

$$m_{x_0} = i\frac{a - a^*}{\sqrt{2}}\,, \tag{7.12}$$

$$m_{y_0} \simeq \frac{a + a^*}{\sqrt{2}} + \frac{a^3 + (a^*)^3 - 3a^*a^2 - 3(a^*)^2 a}{6\sqrt{2}}\,, \tag{7.13}$$

$$m_{z_0} \simeq 1 - a^*a - \frac{a^4 + (a^*)^4 - 2a^*a^3 - 2(a^*)^3 a}{12}\,. \tag{7.14}$$

The following Fourier representation for $a(\boldsymbol{r})$ (and its complex conjugate) will be used:

$$a(\boldsymbol{r}) = \sum_k a_k \exp(i\boldsymbol{k}\boldsymbol{r}_j)\,,$$

$$a_k = \frac{1}{L_y L_z}\int_0^{L_y} dy \int_0^{L_z} dz\, a(\boldsymbol{r}) \exp(-i\boldsymbol{k}\cdot\boldsymbol{r})\,. \tag{7.15}$$

In general, the magnetic energy can be expressed as

$$\mathcal{E} = \mathcal{E}^{(0)} + \mathcal{E}^{(1)} + \mathcal{E}^{(2)} + \mathcal{E}^{(3)} + \mathcal{E}^{(4)} + \ldots\,, \tag{7.16}$$

where the superscript denotes an order in terms of a and a^*.

The zeroth-order energy term is equal to

$$\mathcal{E}^{(0)} = -VK_1\cos^2\theta_0 - VM_sH_0\cos(\theta_H - \theta_0)\,. \tag{7.17}$$

The equilibrium uniformly magnetized state is defined by the condition: $\partial\mathcal{E}^{(0)}/\partial\theta_0 = 0$, which corresponds to

$$H_K\sin 2\theta_0 = 2H_0\sin(\theta_H - \theta_0)\,. \tag{7.18}$$

Here $H_K = 2K_1/M_s$, is the anisotropy field. In order to have a stable stationary state, we need $\partial^2 \mathcal{E}^{(0)}/\partial\theta_0^2 > 0$. The first-order energy term $\mathcal{E}^{(1)} = 0$ at the equilibrium and this condition coincides with (7.18).

The quadratic term has the form

$$\frac{\mathcal{E}^{(2)}}{M_s V/\gamma} = \sum_k \left[\mathcal{A}_k a_k^* a_k + \frac{\mathcal{B}_k}{2}(a_k a_{-k} + a_k^* a_{-k}^*) \right] , \tag{7.19}$$

where

$$\begin{aligned} \mathcal{A}_k &= \gamma \alpha_E k^2 - \frac{\gamma H_K}{2} \sin^2\theta_0 + \gamma H_0 \cos(\theta_H - \theta_0) \\ &\quad + 2\pi\gamma M_s \left\{ [1 - G(k\tau)] \left(\frac{k_{y0}}{k}\right)^2 + G(k\tau) \right\} , \end{aligned} \tag{7.20}$$

$$\begin{aligned} \mathcal{B}_k &= \gamma \alpha_E k^2 - \frac{\gamma H_K}{2} \sin^2\theta_0 \\ &\quad + 2\pi\gamma M_s \left\{ [1 - G(k\tau)] \left(\frac{k_{y0}}{k}\right)^2 - G(k\tau) \right\} , \end{aligned} \tag{7.21}$$

and $\alpha_E \equiv A/M_s$.

Using the following linear canonical transformation

$$a_k = u_k c_k + v_k c_{-k}^* , \quad a_k^* = u_k c_k^* + v_k c_{-k} , \tag{7.22}$$

$$u_k = \sqrt{\frac{\mathcal{A}_k + \omega_k}{2\omega_k}} , \quad v_k = -\frac{\mathcal{B}_k}{|\mathcal{B}_k|}\sqrt{\frac{\mathcal{A}_k - \omega_k}{2\omega_k}} , \tag{7.23}$$

we obtain

$$\mathcal{E}^{(2)} = \frac{M_s V}{\gamma} \sum_k \omega_k c_k^* c_k , \quad \omega_k = \sqrt{\mathcal{A}_k^2 - \mathcal{B}_k^2} . \tag{7.24}$$

The magnon spectrum, ω_k, in an explicit form is

$$\begin{aligned} \omega_k = \gamma &\Bigg\{ H_0 \cos(\theta_H - \theta_0) - H_K \sin^2\theta_0 \\ &+ 4\pi M_s [1 - G(k\tau)] \left(\frac{k_{y0}}{k}\right)^2 + 2\alpha_E k^2 \Bigg\}^{1/2} \\ &\times \left[H_0 \cos(\theta_H - \theta_0) + 4\pi M_s G(k\tau) \right]^{1/2} . \end{aligned} \tag{7.25}$$

7.2.1 Magnon Interactions

The interaction energy can be represented in the form:

$$\frac{\mathcal{E}_{\text{int}}}{M_s V/\gamma} = \sum_{1,2,3}\left[\frac{\Psi_1(1,2,3)}{3}c_1c_2c_3 + \Psi_2(1,2;-3)c_1c_2c_{-3}^* + \text{c.c.}\right]\Delta(1+2+3) + \frac{1}{2}\sum_{1,2,3,4}\Phi(1,2,3,4)c_1^*c_2^*c_3c_4\Delta(1+2-3-4)\,. \tag{7.26}$$

Here for simplicity we use the following notations: $\boldsymbol{k}_1 \equiv \mathbf{1}$, $\boldsymbol{k}_2 \equiv \mathbf{2}$, and so on, for example, $\boldsymbol{k}_1 + \boldsymbol{k}_2 + \boldsymbol{k}_3 \equiv \mathbf{1} + \mathbf{2} + \mathbf{3}$. The three-magnon interaction amplitudes are

$$\Psi_1(1,2,3) = \frac{1}{2}\{\psi(1)(u_1+v_1)(u_2v_3+u_3v_2) + \psi(2)(v_1u_3+v_3u_1)(u_2+v_2) + \psi(3)(v_1u_2+v_2u_1)(u_3+v_3)\}\,, \tag{7.27}$$

and

$$\Psi_2(1,2;3) = \frac{1}{2}\{\psi(1)(u_1+v_1)(u_2u_3+v_2v_3) + \psi(2)(u_2+v_2)(u_1u_3+v_1v_3) + \psi(3)(u_3+v_3)(v_1u_2+v_2u_1)\}\,. \tag{7.28}$$

Here

$$\psi(\boldsymbol{k}) = -\frac{\gamma}{\sqrt{2}}\left(\frac{H_K}{2}\sin 2\theta_0 + 4\pi M_s[1-G(k\tau)]\frac{k_{y0}k_{z0}}{k^2}\right)\,. \tag{7.29}$$

The four-magnon interaction amplitude can be expressed as

$$\Phi(1,2,3,4) = \Phi_0(1,2,3,4) + \Phi_s(1,2,3,4) + \Phi_Q(1,2,3,4)\,, \tag{7.30}$$

where

$$\Phi_0 = \left[\gamma H_0\cos(\theta_H-\theta_0) + \gamma H_K\cos^2\theta_0\right] \times \{u_1u_2v_3v_4 + v_1v_2u_3u_4 - \frac{1}{2}[(u_1u_2+v_1v_2)(u_3v_4+v_3u_4) + (u_1v_2+v_1u_2)(u_3u_4+v_3v_4)]\}\,, \tag{7.31}$$

$$\begin{aligned}\Phi_s = \frac{1}{4}\left[\mathcal{P}(1) + \mathcal{P}(2) + \mathcal{P}(3) + \mathcal{P}(4)\right] \\ \times [(u_1 u_2 - v_1 v_2)(v_3 v_4 - u_3 u_4) \\ - (u_1 v_2 + v_1 u_2)(v_3 v_4 + u_3 u_4) \\ - (u_1 u_2 + v_1 v_2)(v_3 u_4 + u_3 v_4) \\ - \frac{1}{2}(v_1 u_2 + u_1 v_2)(v_3 u_4 - u_3 v_4)] ,\end{aligned} \tag{7.32}$$

$$\begin{aligned}\Phi_Q = [\mathcal{Q}(1+2) + \mathcal{Q}(3+4)](u_1 u_2 u_3 u_4 + v_1 v_2 v_3 v_4) \\ + [\mathcal{Q}(1+4) + \mathcal{Q}(2+3)](u_1 v_2 u_3 v_4 + v_1 u_2 v_3 u_4) \\ + [\mathcal{Q}(1+3) + \mathcal{Q}(2+4)](u_1 v_2 v_3 u_4 + v_1 u_2 u_3 v_4) ,\end{aligned} \tag{7.33}$$

$$\frac{\mathcal{P}(\boldsymbol{k})}{\gamma} \equiv \alpha_E k^2 - \frac{H_K}{2} \sin^2\theta_0 + 2\pi M_s[1 - G(k\tau)]\left(\frac{k_{y0}}{k}\right)^2 , \tag{7.34}$$

and

$$\frac{\mathcal{Q}(\boldsymbol{k})}{\gamma} \equiv \alpha_E k^2 - \frac{H_K}{2} \cos^2\theta_0 + 2\pi M_s[1 - G(k\tau)]\left(\frac{k_{z0}}{k}\right)^2 . \tag{7.35}$$

7.2.2 Effective Four-Magnon Interactions

Unitary transformation (see, Appendix C) makes it possible to eliminate forbidden three-magnon interaction terms in (7.26) and obtain effective interaction terms. As a result we have the following spin-wave energy

$$\begin{aligned}\frac{\mathcal{E}}{M_s V/\gamma} = \sum_k \omega_k c_k^* c_k \\ + \frac{1}{2}\sum_{1,2,3,4} \tilde{\Phi}(1,2,3,4) c_1^* c_2^* c_3 c_4 \Delta(1+2-3-4) ,\end{aligned} \tag{7.36}$$

where

$$\tilde{\Phi}(1,2,3,4) = \Phi(1,2,3,4) + \Phi_1(1,2,3,4) + \Phi_2(1,2,3,4) , \tag{7.37}$$

$$\begin{aligned}\Phi_1(1,2,3,4) = -\Psi_1(1,2,-1-2)\Psi_1(3,4,-3-4) \\ \times\left(\frac{1}{\omega_1 + \omega_2 + \omega_{-1-2}} + \frac{1}{\omega_3 + \omega_4 + \omega_{-3-4}}\right) ,\end{aligned} \tag{7.38}$$

and

$$\begin{aligned}\Phi_2(1,2,3,4) = {} & \Psi_2(1,2,1+2)\Psi_2(3,4,3+4) \\ & \times \left(\frac{1}{\omega_1+\omega_2-\omega_{1+2}} + \frac{1}{\omega_3+\omega_4-\omega_{3+4}}\right) \\ & - \Psi_2(1,3-1,3)\Psi_2(4,2-4,2) \\ & \times \left(\frac{1}{\omega_1+\omega_{3\text{-}1}-\omega_3} + \frac{1}{\omega_4+\omega_{2-4}-\omega_2}\right) \\ & - \Psi_2(1,4-1,4)\Psi_2(3,2-3,2) \\ & \times \left(\frac{1}{\omega_1+\omega_{4\text{-}1}-\omega_4} + \frac{1}{\omega_3+\omega_{2-3}-\omega_2}\right) \\ & - \Psi_2(2,3-2,3)\Psi_2(4,1-4,1) \\ & \times \left(\frac{1}{\omega_2+\omega_{3-2}-\omega_3} + \frac{1}{\omega_4+\omega_{1-4}-\omega_1}\right) \\ & - \Psi_2(2,4-2,4)\Psi_2(3,1-3,1) \\ & \times \left(\frac{1}{\omega_2+\omega_{4-2}-\omega_4} + \frac{1}{\omega_3+\omega_{1-3}-\omega_1}\right) . \end{aligned} \tag{7.39}$$

The energy (7.36) together with Hamilton's equations of motion for complex spin-wave variables

$$\left(\frac{d}{dt} + \eta_k\right) c_k = -i\left(\frac{\gamma}{M_s V}\right)\frac{\partial \mathcal{E}}{\partial c_k^*} , \tag{7.40}$$

supplemented by the (microscopic) relaxation rate η_k, represent the basis for magnetization dynamics modeling in $\boldsymbol{k}$-space. Calculating $c_k(t)$, $c_k^*(t)$, we can find magnetization deviations $a_k(t)$, $a_k^*(t)$ with the help of back transformation (7.22) and finally with (7.12)–(7.14), the averaged

$$\langle \boldsymbol{m}(\boldsymbol{r},t)\rangle = \frac{1}{L_y L_z}\int_0^{L_y} dy \int_0^{L_z} dz\, \boldsymbol{m}(\boldsymbol{r},t) = \boldsymbol{m}_0(t) , \tag{7.41}$$

which gives a measure of nonuniform magnetization motions in the system. In general, $|\boldsymbol{m}_0| < 1$ and only in the case of coherent spin motion $|\boldsymbol{m}_0| = 1$.

7.3 Example

For simplicity we shall consider the uniform magnetization precession interacting with one magnon pair with $\boldsymbol{k} = (0,0,k)$ along $\hat{z}_0$. In this case the energy (Hamiltonian) of the system can be reduced to the form:

$$\mathcal{H} = \mathcal{H}_0 + \mathcal{H}_{\text{int}} + \mathcal{H}_p , \tag{7.42}$$

where

$$\frac{\mathcal{H}_0}{M_s V/\gamma} = \omega_0 c_0^* c_0 + \omega_k (c_k^* c_k + c_{-k}^* c_{-k}) \tag{7.43}$$

describes the uniform precession and magnon pair,

$$\begin{aligned}\frac{\mathcal{H}_{\text{int}}}{M_s V/\gamma} &= \frac{\Phi_{00}}{2} c_0^* c_0^* c_0 c_0 + 2\Phi_{0k} c_0^* c_0 (c_k^* c_k + c_{-k}^* c_{-k}) \\ &+ \frac{\Phi_{kk}}{2} (c_k^* c_k c_k^* c_k + c_{-k}^* c_{-k} c_{-k}^* c_{-k}) + 2\Phi_{k,-k} c_k^* c_k c_{-k}^* c_{-k}\end{aligned} \tag{7.44}$$

describes nonlinear interactions in the system,

$$\Phi_{00} \equiv \tilde{\Phi}(0,0,0,0)\,, \quad \Phi_{0k} \equiv \tilde{\Phi}(\boldsymbol{k},0,\boldsymbol{k},0)\,,$$

$$\Phi_{kk} \equiv \tilde{\Phi}(\boldsymbol{k},\boldsymbol{k},\boldsymbol{k},\boldsymbol{k})\,, \quad \Phi_{k,-k} \equiv \tilde{\Phi}(\boldsymbol{k},-\boldsymbol{k},\boldsymbol{k},-\boldsymbol{k})\,,$$

and

$$\frac{\mathcal{H}_p}{M_s V/\gamma} = \Phi_p \left(c_0 c_0 c_k^* c_{-k}^* + c_0^* c_0^* c_k c_{-k} \right) \tag{7.45}$$

describes the spin wave-pair excitation by the uniform precession, $\Phi_p \equiv \tilde{\Phi}(\boldsymbol{k},-\boldsymbol{k},0,0)$.

The uniform precession and magnon-pair dynamics are defined by

$$\left(\frac{d}{dt} + \eta_0 \right) c_0 = -i\tilde{\omega}_0 c_0 - 2i\Phi_p c_k c_{-k} c_0^*\,, \tag{7.46}$$

$$\left(\frac{d}{dt} + \eta_k \right) c_k = -i\tilde{\omega}_k c_k - i\Phi_p (c_0)^2 c_{-k}^*\,, \tag{7.47}$$

where

$$\tilde{\omega}_0 = \omega_0 + \tilde{\Phi}_{00}|c_0|^2 + 2\Phi_{0k}(|c_k|^2 + |c_{-k}|^2)\,, \tag{7.48}$$

$$\tilde{\omega}_k = \omega_k + 2\tilde{\Phi}_{0k}|c_0|^2 + \Phi_{kk}|c_k|^2 + 2\Phi_{k,-k}|c_{-k}|^2 \tag{7.49}$$

are nonlinear frequencies and η_0, η_k are the relaxation rates.

From the energy symmetry we have $c_k = c_{-k}$. Thus, (7.46)–(7.49) represent a self-consistent nonlinear theory of magnetization reversal with just two independent complex variables. Simple analysis for both $c_k(t)$ and $c_{-k}^*(t) \propto \exp(\kappa t)$, where κ is an increment of instability, gives:

$$\kappa = -\eta_k + \sqrt{\left|\Phi_p c_0^2(0)\right|^2 - (\tilde{\omega}_k - \tilde{\omega}_0)^2}\,. \tag{7.50}$$

This formula is similar to that obtained above (4.15) for parametric instability. The difference is that the resonance condition includes nonlinear frequencies:

$2\tilde{\omega}_0 = \tilde{\omega}_k + \tilde{\omega}_{-k}$ (all magnons out of this equality can be included into a thermal bath) and $\tilde{\omega}_0$ is the uniform precession frequency (not a driving field frequency). The onset of instability is defined by $|\Phi_p c_0^2(0)| \geq \eta_k$. Taking $m_{x0}(0) = 0$, from (7.10), (7.11) and (7.22) we can find $c_0(0)$ and rewrite this criterion as

$$\theta \geq (u_0 + v_0)\sqrt{\frac{2\eta_k}{|\Phi_p|}}\,, \tag{7.51}$$

where $\theta \equiv \tan^{-1}(m_{y0}/m_{z0})$ is the initial deviation angle from the equilibrium direction z_0. Here the parameters u_0, v_0 and Φ_p can be directly calculated using (7.22), (7.45). The damping η_k can be estimated microscopically (see, Chapter 3).

Figure 7.2a, b shows two different evolutions of the magnetic system calculated by (7.46), (7.47). Magnon-pair excitation in Figure 7.2a is not sufficiently strong during the switching process to affect the uniform magnetization dynamics. The averaged $\langle \boldsymbol{m}(\boldsymbol{r})\rangle = \boldsymbol{m}_0$, which gives a measure of nonuniform magnetization motions in the system is relatively small ($1 - |\boldsymbol{m}_0| \lesssim 0.05$). Figure 7.2b demonstrates a strong excitation of magnon instability by uniform precession and substantial increase of the magnetization reversal rate. In this case $|\boldsymbol{m}_0|$ reaches $\simeq 0.6$. For stronger coupling ($|\Phi_p|/\eta_k > 21$) we observed beating between the uniform precession and the magnon pair. In this case it is necessary to consider the excitation of another resonance magnon pair.

Thus, we have considered an ultrathin ferromagnetic film with large dimensions in the plane. In this case the role of plane boundaries is negligible and the most

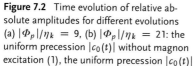

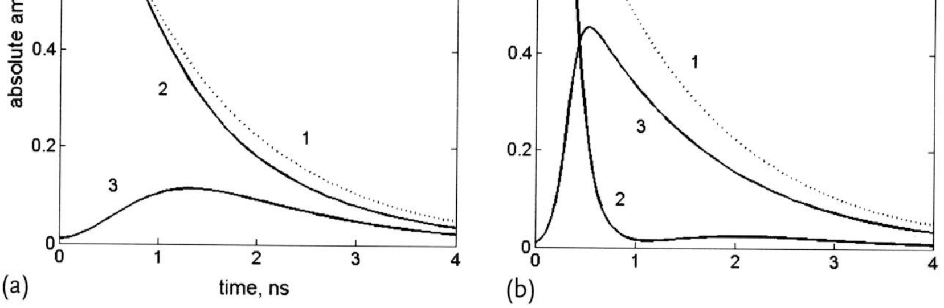

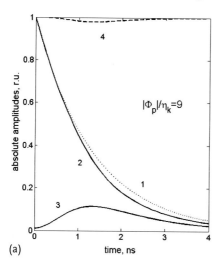

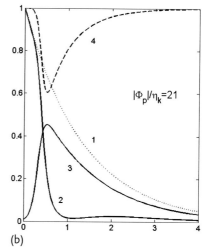

Figure 7.2 Time evolution of relative absolute amplitudes for different evolutions (a) $|\Phi_p|/\eta_k = 9$, (b) $|\Phi_p|/\eta_k = 21$: the uniform precession $|c_0(t)|$ without magnon excitation (1), the uniform precession $|c_0(t)|$ with magnon excitation (2), excited magnons $|c_k(t)|$ (3). Curve (4) describes $|\boldsymbol{m}_0(t)|$. The experimental conditions correspond to Figure 2 in [143].

convenient technique to describe the nonuniform spin motions is their magnon representation in $\boldsymbol{k}$-space. Taking into account linear magnon modes and their scattering, we have constructed a nonlinear self-consistent theory of magnetization reversal as a decay of uniform magnetization precession and nonlinear excitation of magnon pairs. This theory includes an effective energy (7.36) and dynamic (7.40). The most important magnon modes are defined by the resonance condition $2\tilde{\omega}_0 = \tilde{\omega}_k + \tilde{\omega}_{-k}$ (similar to Suhl's second-order instability). The excitation of all magnons out of this resonance is small and therefore they can be considered as part of a thermal bath. In the simple example presented above, we have demonstrated that a strongly excited magnon instability can substantially increase the magnetization switching rate.

7.4 Discussion

a) What is micromagnetic modeling in $\boldsymbol{k}$-space?
b) Why do we need to eliminate ineffective three-magnon terms?
c) Why is the Holstein–Primakoff transformation not suitable for two-dimensional systems, and is it better to use other (Villain–Baryakhtar–Yablonskii) transformations of spin components to harmonic oscillator variables?

8 Collective Magnetic Dynamics in Nanoparticles

The rapid progress in nanomagnetic technologies has stimulated a search for new physics and applications of collective spin dynamics [157–159]. Nanomagnetic particles and clusters of magnetic atoms and ions (for example, sub-nanomaganetic objects) are extremely attractive for construction of nanomagnetic devices. In this chapter we will discuss collective quantum dynamics of clusters of localized electronic spins being used as a platform for very efficient analog data-processing devices. This expectation is based on preliminary studies and the analogy with clusters of dipolar-coupled nuclear spins, where this type of parallel information processing has been experimentally demonstrated.

For clusters of coupled nuclear spins with conventional unresolved NMR spectra, it has been found that it is possible to excite long-lived coherent response signals. With multifrequency excitation, one can store and process in parallel more than a kilobit of information [160–162]. An example of efficient parallel data processing has been achieved by implementing the parallel search algorithm [163, 164].

For efficient information processing a cluster should be large enough to have complicated dynamics but small enough to remain in the quantum scale. The physical system which was used in [165] is the nematic liquid crystal 4-*n*-pentyl-4′-cyanobiphenyl (5CB). Fast molecular motions average interactions between molecules, but 19 proton spins of one molecule are coupled with residual dipole–dipole interactions. Therefore, the system is a good model of an ensemble of noninteracting spin clusters.

An example of parallel logic operation with a 64-bit array is shown in Figure 8.1. It is an implementation of a parallel bitwise NOT operation, which flips the bits by changing ones to zeros, and vice versa. The spectrum in Figure 8.1c represents the result of a parallel bitwise NOT operation simultaneously applied to each bit of the spectrum in Figure 8.1b. It was obtained when a pulse similar to that used for Figure 8.1b was applied in antiphase after a pulse which was used to generate Figure 8.1a, so that the corresponding peaks in Figure 8.1a are "erased".

Let us consider a hypothetical device with parallel information processing using long-lived collective spin excitations with different frequencies in nanomagnetic particles. The device has relatively simple classical computers – classical input and output. On the other hand, it has a power of quantum computations with magnetic clusters, high efficiency of information processing, and low heating. A signal, con-

Nonequilibrium Magnons, First Edition. Vladimir L. Safonov.

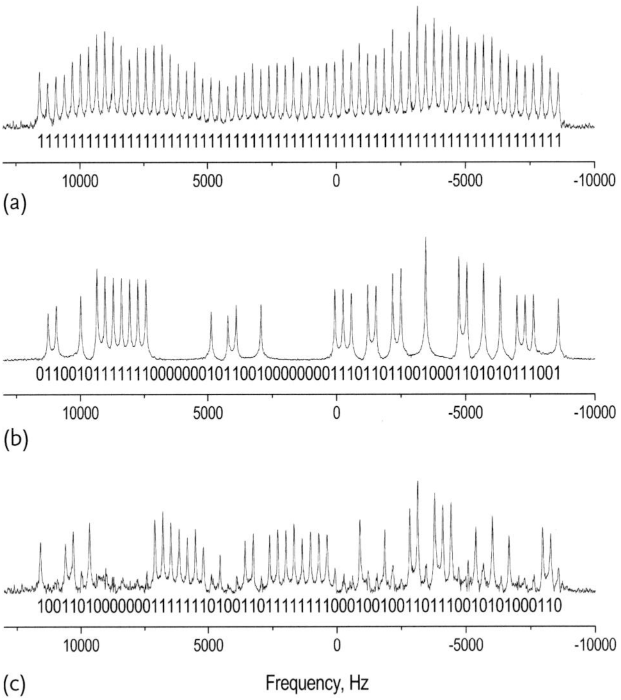

Figure 8.1 A parallel bitwise NOT gate operation using a spin cluster of the nematic liquid crystal 4-*n*-pentyl-4′-cyanobiphenyl (5CB). (a) Spectrum of the absolute value spectrum generated by a 50 ms pulse which is the sum of 64 harmonics. It represents the number $2^{64} - 1$ in binary notation. (b) Spectrum of the spectral representation of an arbitrarily chosen integer number $0 \leq x < 2^64$ obtained by applying a different 50 ms 64-frequency pulse, with some of the amplitudes set to zero as indicated below the spectrum. In decimal notation, $x = 7\,348\,754\,808\,244\,345\,529$. (c) Spectrum obtained by the pulse in (a) followed by a 10 ms pulse similar to that used for spectrum (b), applied in antiphase. It represents the number $y = 2^{64} - 1 - x$, or $y = 11\,097\,989\,265\,465\,206\,086$ in decimal notation. The amplitude of the RF field per harmonic was 10.5 Hz for 50 ms pulses and 2.9 Hz for the 10 ms pulse. The number of transients was 1024 [165].

taining a spectrum of frequencies or a sequence of microwave pulses (1–100 GHz range, each pulse can encode a bit of information) excites, through a microcoil, collective magnetic dynamics in magnetic clusters. The second signal or the second sequence of pulses interacts with the magnetic oscillations and changes the collective dynamics. The response from the magnetic system, received by the same coil, gives the result of calculation. The parallel processor has a processing speed up to 10^{11} bit/s at room temperature and works in the 1–100 GHz region. It can utilize both the binary, ternary and higher-order (fuzzy) logics. Parallel spin processor utilizing parallel computations (about 100 bits or more) can be incorporated as coprocessor in a hybrid scheme with conventional digital electronics. This combined analog–digital processor can be used to significantly enhance the performance of a standard PC and reduce power consumption. The parallel spin processing of information can also be used as a good basis for neural networks – the next generation of computer systems.

The following lists the types of advantages we can expect:

Simultaneous parallel operations One advantage lies in the manipulations of the collective spin states of a composite quantum system. As a result it is not necessary to use "qubits" or quantum logic gates as a basis of operations. It is possible to "unfold" the internal complexity of the spin system into a long array of classical bits (collective spin excitations with different frequencies), which can be manipulated in parallel.

Classical input and output Parallel processor has relatively simple classical computers with the ability to utilize classical input and output technology. On the other hand, it has a power of quantum computations through the utilization of collective quantum dynamics.

Small power consumption A natural amplification provided by the collective spin dynamics of a spin cluster requires much less power than that used in conventional digital processing for the same processing speed.

Negligible thermal effect The parallel spin processor implements dynamic information processing with negligible thermal effects. In the simplest application we can replace a standard binary processor containing, say, 100 transistors (which produce heat) with just one spin processor (where heat production is negligible) working with 100 binary operations in parallel.

Stacking technology Stacking (third dimension) technology is used for the magnetic processor versus planar semiconductor technology.
Logic devices with nanomagnetic particles will be stable against radiation and can work at much higher temperatures compared to semiconductor crystals. This may be useful in a device that orbits outside the Earth's atmosphere and is subject to extreme heat variation and direct radiation.

Binary and Ternary numbers and logics The parallel processor can utilize both the binary and ternary numbers and logics.

The above parallel spin processor can consist of a set of nanomagnetic particles (containing transition or rare-earth elements) with a processing speed up to 10^{11} bit/s at room temperature. Examples of nanomagnetic particles are $FeBO_3$, FeF_3, MnTe, MnO (Mn^{2+}, Fe^{3+}, Eu^{2+}, Gd^{3+} are in an S state with zero orbital momentum). It is no challenge to construct a micron-sized parallel processor; however, it is still difficult when constructing sizes approaching tens of nanometers, depending on the size of microresonator or microcoil and sensitivity of the conventional electronics coupled to the processor. The input and output signals of such a device are classical (1–100 GHz region). The processor with parallel computations (about 100 bits or more) can be incorporated, as co-processor, in hybrid schemes using conventional digital electronics. In addition to the large number of simultaneous parallel operations, the spin processor can implement dynamic information processing with negligible thermal effects.

For parallel bitwise operations, the maximum total speed of data processing is equal to the strength of internal spin–spin interactions (in frequency units) or, more precisely, to the width of the conventional linear response spectrum.

The parallel processor sufficiently increases the rate of information processing and presents the possibility to work with huge arrays of data and real-time data streams. The parallel processors can be used, for example, in video cards to sufficiently enhance the graphic traffic.

There are several reasons why the new type of information processing can be both feasible and efficient.

At present, analog devices cannot compete with digital ones due to accumulation of errors and the very low informational content of analog signals. The informational content of the multifrequency coherent response signals can be very large (it exceeded one kilobit in the experiments with a system of 19 nuclear spins [160]). The problem of errors can be solved by combining such analog devices with conventional digital electronics, wherein fast computation is performed in an analog mode and then the result is digitized. The information can be encoded both in the frequency domain (by turning on and off the amplitudes of harmonics) and the time domain (as a sequence of microwave or magnetic field pulses).

The process of collective spin excitation is very nonlinear and, therefore, makes it possible to implement parallel bitwise logic operations with long bit strings [163].

A physical system with two stable states, representing a digital bit, requires some time for equilibration and energy dissipation after each switching. Dynamical (analog) systems can operate much faster and are more energy-efficient. With parallel processing, information can be processed much faster than flips of individual spins. The total power required, for the same processing speed, is inversely proportional to the number of bits which can be manipulated in parallel.

As it is shown in the next section, the polarization of spins with external magnetic field is not necessary. Partial order, created by internal interactions, is sufficient. Therefore, the device can be just a small sample inside a microcoil and will not need an external magnet. Compared to nuclear spins, electronic clusters should produce about ten orders of magnitude stronger signals, and samples of a submicron size produce sufficiently strong signals.

8.1
Long-Lived States in a Cluster of Coupled Nuclear Spins

All material in this section is the result of discussions with Prof. A.K. Khitrin and is reproduced by his kind permission.

The dynamical complexity of spin clusters increases quickly with the number of coupled spins. For N spins 1/2, coupled by interactions, there are up to $C_{2N}^{N+1} \sim 2^{2N}$ allowable single quantum transitions between $2N$ quantum levels. For a composite classical system the degrees of freedom of its parts are additive, while for a composite quantum system they are multiplicative. Practically, this means that computer simulations of spin dynamics in clusters with more than a dozen coupled spins 1/2 become very difficult tasks. When the dynamics is too complicated for detailed analysis, a thermodynamic approach based on the concept of spin temperature [97] is often used. The most successful example is Provotorov's saturation theory [166], which describes interaction with a weak excitation field and explains various phenomena in solid-state NMR. The spin-temperature approach is based on the assumption that if, in an initial state, a system with the Hamiltonian $\mathcal{H}_0$ deviates from equilibrium, it evolves to the equilibrium state with

$$\rho_0 = Z - \mathbf{1} \cdot \exp(-\beta \mathcal{H}_0) , \tag{8.1}$$

where $Z = \mathrm{Tr}[\exp(-\beta \mathcal{H}_0)]$, ρ_0 is the equilibrium density matrix, $\beta = 1/k_\mathrm{B} T$ and T is the equilibrium spin temperature. The characteristic time of this evolution is the characteristic time of spin–spin interactions.

We will demonstrate that in the absence of spin-lattice relaxation it is possible to excite arbitrarily long response signals and tailor any shapes of the corresponding response spectra. This means that the informational content of the response signal can be very high.

Let us consider the high-temperature initial state with the density matrix

$$\rho_\mathrm{in} = \frac{\mathbf{1} - (\beta/2) \int d\omega\, f(\omega)(S_\omega + S_{-\omega})}{\mathrm{Tr}\mathbf{1}} , \tag{8.2}$$

where $f(\omega)$ is some arbitrary function of ω, S_ω is the Fourier component (in the interaction representation) of $S_x(t)$, and S_x is the x-component of the total spin:

$$S_x = \int d\omega\, S_\omega , \tag{8.3}$$

$$S_x(t) = \exp(i\mathcal{H}_0 t)\, S x \exp(-i\mathcal{H}_0 t) = \int d\omega\, S_\omega \exp(i\omega t) . \tag{8.4}$$

The following commutation relation takes place:

$$[\mathcal{H}_0, S_\omega] = \omega S_\omega . \tag{8.5}$$

At $f(\omega) \equiv 1$ the initial state (8.2) corresponds to a uniform polarization along the x-axis:

$$\rho_\mathrm{in} = \frac{\mathbf{1} - \beta S_x}{\mathrm{Tr}\mathbf{1}} . \tag{8.6}$$

If the observable is S_x, for the initial condition (8.2) one obtains the following expression for the response signal:

$$\begin{aligned}
\langle S_x(t) \rangle &= \mathrm{Tr}\left[S_x(t)\rho_\mathrm{in}\right] \\
&= 2^{-N}\left(\frac{\beta}{2}\right) \int\int d\omega\, d\omega'\, f(\omega) \exp(i\omega' t) \mathrm{Tr}\left[(S_\omega + S_{-\omega}) S_{\omega'}\right] \\
&= 2^{-N}\beta \int d\omega\, f(\omega) \cos(\omega t) \mathrm{Tr}(S_\omega S_{-\omega}) \\
&= \beta \int d\omega\, f(\omega) g(\omega) \cos(\omega t) ,
\end{aligned} \tag{8.7}$$

where $g(\omega) = 2^{-N}\mathrm{Tr}(S_\omega S_{-\omega})$ is the conventional linear response spectrum. In (8.7) the following property of the traces has been used:

$$\mathrm{Tr}(S_\omega S_{\omega'}) = \delta(\omega + \omega')\mathrm{Tr}(S_\omega S_{-\omega}) \,. \tag{8.8}$$

We see that, by choosing the function $f(\omega)$ in the initial state (8.2), one can obtain any response spectrum within the shape $g(\omega)$. As an example, when $f(\omega)$ is the δ-function, the response signal (8.7) oscillates without decay and the corresponding spectrum also has a shape of the δ-function.

This calculation shows that it is possible to excite sharp response signals in large systems of coupled spins. The ability to create a large number of distinguishable shapes of the response signal means that the "output" signal can carry a large amount of information.

Collective coherent response signals of spin clusters make it possible to "communicate" with a mesoscopic quantum system, bring a large amount of information to the system, process it in parallel, and get an output signal with high informational content. This type of collective response is not just a conversion of the existing spin polarization into a detectable signal, as it would be in the case of inhomogeneously broadened spectra. The signal can also be easily excited when the initial spin state of the cluster corresponds to dipolar ordering and there is no overall magnetization in the initial state.

Experimental results and computer simulations suggest that the new type of collective response is a general feature of mesoscopic quantum systems. For practical applications, this means that prospective devices do not need external magnets to create signals if the internal spin–spin interactions are sufficiently strong.

8.2
Electronic Spins

Electronic spin has about a 1836 times larger magnetic moment than the spin of a proton. Theoretically, it is much easier to excite and detect collective spin cluster dynamics (mentioned above for nuclear spins) in electronic spin clusters. Typical frequencies of this electronic spin collective dynamics depend on the anisotropic spin–spin interaction strength, and for ions of transition and rare-earth elements they are in the $1{-}10^4$ GHz range. For example, the dipole–dipole interactions between electronic spins are, at the same distances, $\sim 2000^2 = 4 \times 10^6$ times (more than six orders of magnitude!) larger than for nuclear spins and their collective spin dynamics can reach frequencies 10–100 GHz. For ferromagnetic (or antiferromagnetic) material with anisotropic exchange interaction strength $|J|/k_\mathrm{B} \sim 100$ K, the corresponding collective spin dynamics frequency in the paramagnetic state is on the order of ~ 2 THz. Another type of anisotropic interaction, which can act favorably, is a single-spin g-factor anisotropy, if it is different for different spins of the cluster. Our goal is to search for a combination of interactions suitable for producing coherent long-lived response signals in a broad frequency range, and additionally we aim to propose some realistic systems for further study. For parallel bitwise operations, the maximum total speed of data processing is equal to the strength of internal spin–spin interactions (in frequency units) or, more precisely, to the width of the conventional linear response spectrum.

The electronic spin clusters have several characteristic features. First, the electronic spins are usually greater than 1/2, for example, Cu^{2+}, V^{4+}, Yb^{3+} have $S = 1/2$, Ni^{2+}, Tm^{3+} have $S = 1$, Co^{2+}, Mn^{4+}, Er^{3+} have $S = 3/2$, Cr^{2+}, Mn^{3+}, Ho^{3+} have $S = 2$, and so on. Second, only Mn^{2+}, Fe^{3+} ($S = 5/2$) and Eu^{2+}, Gd^{3+} ($S = 7/2$) are in an S state with zero orbital momentum. Other ions of transition and rare-earth elements have nonzero orbital momentum and corresponding spin-orbital coupling. Third, in addition to the dipole–dipole and Zeeman energies (which are typical for the nuclear spin systems) the electronic spin system Hamiltonian usually includes strong ferro- or antiferromagnetic exchange interactions and various energies of single-spin and two-spin anisotropy. There are a number of electronic spin systems that are characterized by different values of electronic spin, presence of orbital magnetism, and signs of the exchange and anisotropy energies. Fourth, the electronic spin cluster can be in a molecular form, in a form of nano- and sub-nanoparticle (of magnetic and paramagnetic material), on the surface and in the bulk of material. All these factors must be taken into account for a detailed classification scheme of electronic spin clusters.

Another important characteristic of a spin system is the requirement that the spin-lattice relaxation remains relatively small. Spin-lattice relaxation and thermal fluctuations from the thermal reservoir destroy Hamiltonian behavior of the spin system and would also destroy the coherent long-lived signals. Electronic spin interactions with the lattice (thermal reservoir) are much stronger than that for nuclear spins and there is a large variety of them. The elastic modulation of exchange interaction and crystalline fields creates a coupling to phonons; spin–orbital interactions contribute to so-called fast and slow spin-lattice relaxation mechanisms. However, the spin-lattice relaxation of electronic spins can be surprisingly slow, especially for dielectric systems. As an example, the quality factor of ferromagnetic resonance in YIG (yttrium iron garnet) can reach 10^6. The corresponding factor (the ratio of internal interactions to the spin-lattice relaxation rate) for the studied nuclear spin clusters was only about 10^4.

For magnetic nanoparticles the thermal reservoir, which absorbs the energy of excitations becomes very small. The dipole–dipole interaction energy between the two nearest cells of the linear size L with the magnetization M decreases as

$$\mathcal{E}_{\text{dd}} \sim \frac{(M_1 L^3)(M_2 L^3)}{L^3} \sim M^2 L^3 .$$

Taking into account typical sound velocity $v_s \sim 10^3$ m/s and a linear size of nanoparticle $d \sim 1$ nm, one can also obtain the following estimate for the lowest frequency of elastic vibrations of the sample:

$$f \sim v_s/(d) \sim 1\,\text{THz} .$$

The role of KAM theorem in the process of magnetic relaxation of small magnetic particles has been discussed in [145, 167].

8.3
Spin-Echo Logic Operations

Spin echo is a very useful and practical phenomenon. The spin-echo pulse sequence is commonly used in nuclear and electronic magnetic resonance.

Spin-echo memory on nuclear spins was proposed independently by Fernbach and Proctor [168], and Anderson, Garwin, Hahn *et al.* [169] in 1955 (see above). A sequence of RF pulses (code) is applied to the sample and after the reading pulse there is spin-echo sequence (in a reverse order). Applying an extra starting pulse, one can observe the spin-echo as a proper code. Figure 8.2 illustrates an example of the sequence of microwave pulses and spin echo.

It should be noted that the phenomenon of spin echo was observed in the system of electronic spins [170] and in systems of parametrically pumped magnons [171–173]. As it was recently shown, the echo effect is a common phenomenon in coupled spin systems [174, 175].

In this final section we shall discuss a method to perform logic operations using microwave pulses and spin echo in paramagnetic materials.

Microwave pulses from independent sources are applied to the paramagnetic sample in a resonator (coil) by antennas. A superposition of two (or more) sequences of microwave pulses (strings) excites electronic paramagnetic resonance dynamics in the paramagnetic sample. The response (spin-echo) from the paramagnetic system being received by the same resonator (coil) gives the result of the calculation.

The elementary logic operation consists of the four steps: (i) A starting microwave pulse is applied to get stimulated echo in a proper sequence. (ii) Two independent synchronized sequences of pulses (two binary strings) are applied to the sample. The resulting signal excites the spin resonance dynamics. (iii) A reading microwave pulse is applied to recall a code. (iv) A recall code as the echo sequence is registered. The reading sensor detects the echo signals as a result of the logic operation *OR*, *AND*, *XOR* or *NOT*.

For logic operations with spin echo we represent codes using 1 (pulse), 2 (pulse with two times larger amplitude) and 0 (no pulse). For example, the code 2011 is shown as signal C in Figure 8.3.

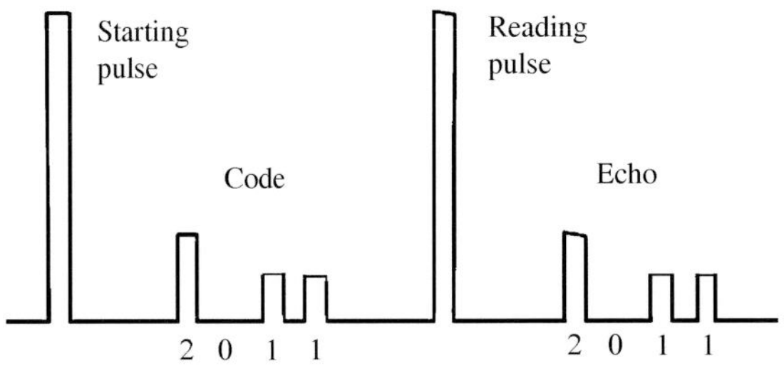

Figure 8.2 Sequence of events with starting and reading pulses, code and corresponding spin echo.

For an echo detector that detects any pulse

$$C = 1011 = (1010)\mathrm{OR}(1001) = (A)\mathrm{OR}(B)\,.$$

For an echo detector that detects only the largest pulses we have

$$D = 1000 = (1010)\mathrm{AND}(1001) = (A)\mathrm{AND}(B)\,.$$

Figure 8.3 illustrates the OR operation with a simple echo detector (that does not distinguish the height of pulses):

$$(1010)\mathrm{OR}(1001) = 1011\,.$$

For an advanced echo detector that detects only large pulses we have the AND logical operation:

$$(1010)\mathrm{AND}(1001) = 1000\,.$$

One can introduce a concept of a "negative" (−1) pulse. This pulse has 180° phase shift with respect to the 1 pulse. There are the following simple relations:

$$(-1) + (-1) = (-2), (-1) + (1) = (1) + (-1) = 0\,.$$

Here the parentheses are shown for simplicity of representation. The negative pulse will be represented as a shaded pulse. It is easy to check that superposition of B and −B cancel each other.

Figure 8.4 illustrates the XOR operation. Superposing the first number A = 1010 and the second number B (in the form of −B = −100−1), we obtain C = 001−1, or, neglecting the sign, C = 0011. In other words, (A)XOR(B) = C.

Figure 8.5 illustrates the NOT operation. The string 1010 is superposed with the sequence of negative pulses −1−1−1−. Neglecting the sign of resulting pulses, we obtain:

$$\mathrm{NOT}(1010) = 0101\,.$$

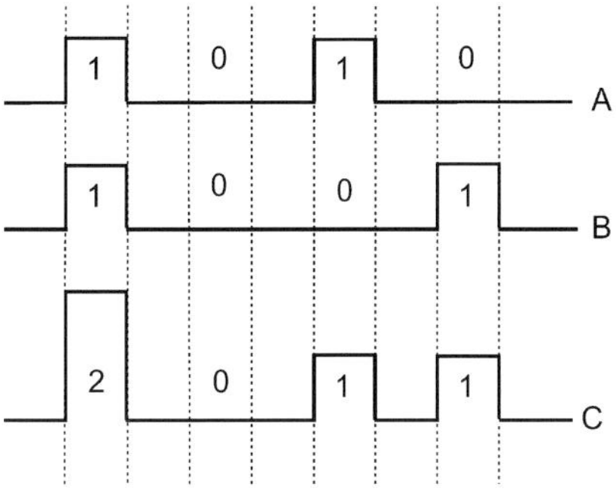

Figure 8.3 The first string A = 1010. The second string B = 1001. The result of superposition is C = 2011. The vertical dotted lines separate different bits.

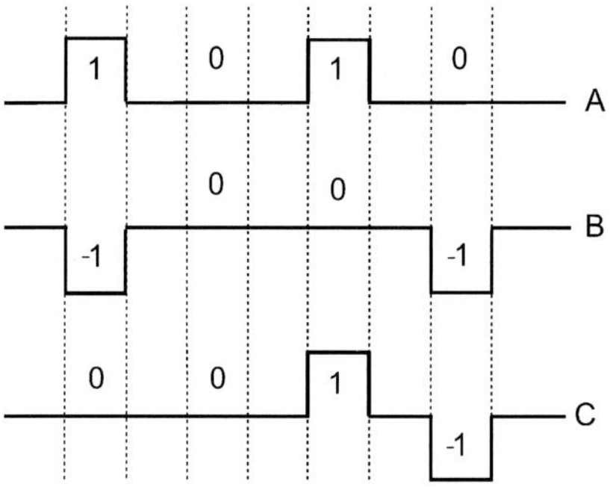

Figure 8.4 A = 1010, −B = −100−1. The result of superposition is C = 001−1. So far as the sign of the pulse in C does not matter for a detector, C = 0011 = (1010)XOR(1001) = (A)XOR(B).

The following are characteristics of a device with spin-echo logic operations:

- It will have negligibly small heating effect. Heating is due to spin-lattice relaxation only.
- Stacking (third dimension) technology can be used for the spin-echo logic device versus planar semiconductor technology.
- The spin-echo logic device can utilize both binary and ternary numbers and logics.
- It will be stable to X-ray radiation and can work at much higher temperatures compared to semiconductor crystals.
- It can be incorporated in hybrid schemes using conventional digital electronics.

The results of A1 = (A)XOR(B) and C1 = (A)AND(B) logic operations can be used to construct an adder. We represent B1 = LeftShift(C1), where LeftShift(.) denotes the left shifted C1 by one register, and find A2 = (A1)XOR(B1) and C2 = (A1)AND(B1). Then B2 = LeftShift(C2) and all procedures are repeated n times until Cn is equal to zero in all registers. The final result is An. Analogously, other

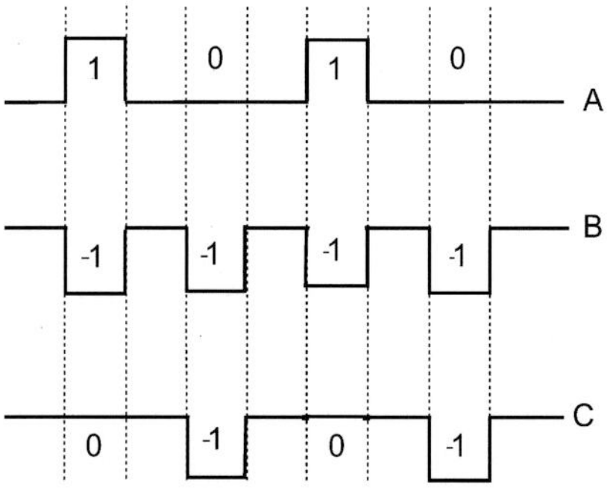

Figure 8.5 A = 1010. B = −1−1−1−1. The superposition C = 0−10−1. As far as the sign of the pulse in C does not matter for a detector, we have C = NOT(A) = 0101.

arithmetic operations can be developed. Therefore, using this logic one can organize a complete data processor. The logic operations described above can be developed for other echo systems and systems with superimposition.

Appendix A
Harmonic Oscillator in Quantum Mechanics

Here, we shall briefly discuss the basic quantum-statistical properties of harmonic oscillators. A more detailed description can be found, for example, in [5, 46, 176, 177].

The quantum Hamiltonian of a harmonic oscillator can be written in the form:

$$\hat{\mathcal{H}} = \frac{\hat{p}^2}{2m} + \frac{m\omega^2\hat{x}^2}{2} \tag{A1}$$

where $\hat{p} = -i\hbar(d/dq)$ is the operator of momentum, $\hat{x} = x$ is a coordinate, m is the mass and ω is the frequency of harmonic oscillations.

The commutation relation for $\hat{p}$ and $\hat{x}$ operators are

$$[\hat{x}, \hat{p}] \equiv \hat{x}\hat{p} - \hat{p}\hat{x} = i\hbar\,. \tag{A2}$$

The Schrödinger equation for a harmonic oscillator can be written as

$$\hat{\mathcal{H}}|n\rangle = \epsilon_n|n\rangle\,, \tag{A3}$$

where

$$\epsilon_n = \hbar\omega\left(n + \frac{1}{2}\right)\,, \qquad n = 0, 1, 2, \ldots$$

are the eigenvalues (energy spectrum) of operator $\hat{\mathcal{H}}$ and $|n\rangle$ represent the corresponding eigenfunctions.

A.1
Operators of Creation and Annihilation

The creation and annihilation operators are defined by

$$\begin{aligned} \hat{a}^\dagger &= \frac{1}{2}\left(\frac{\hat{x}}{x_0} - i\frac{\hat{p}}{p_0}\right)\,, \\ \hat{a} &= \frac{1}{2}\left(\frac{\hat{x}}{x_0} + i\frac{\hat{p}}{p_0}\right)\,, \end{aligned} \tag{A4}$$

Nonequilibrium Magnons, First Edition. Vladimir L. Safonov.

where

$$x_0 \equiv \sqrt{\frac{\hbar}{2m\omega}}\,, \quad p_0 \equiv \sqrt{\frac{m\hbar\omega}{2}}\,. \tag{A5}$$

The creation and annihilation operators have the following mathematical properties:

$$[\hat{a}, \hat{a}^\dagger] \equiv \hat{a}\hat{a}^\dagger - \hat{a}^\dagger\hat{a} = 1\,, \tag{A6}$$

and

$$\begin{aligned} \hat{a}|n\rangle &= \sqrt{n}|n-1\rangle\,, \\ \hat{a}^\dagger|n\rangle &= \sqrt{n+1}|n+1\rangle\,. \end{aligned} \tag{A7}$$

The Hamiltonian (A1) in terms of new variables becomes

$$\hat{\mathcal{H}} = \hbar\omega\left(\hat{N} + \frac{1}{2}\right)\,, \quad \hat{N} = \hat{a}^\dagger\hat{a}\,. \tag{A8}$$

The Heisenberg equations of motion for $\hat{a}^+$ and $\hat{a}$ have the form:

$$\begin{aligned} i\frac{d}{dt}\hat{a} &= \left[\hat{a}, \frac{\hat{\mathcal{H}}}{\hbar}\right] = \omega\hat{a}\,, \\ i\frac{d}{dt}\hat{a}^\dagger &= \left[\hat{a}^\dagger, \frac{\hat{\mathcal{H}}}{\hbar}\right] = -\omega\hat{a}^\dagger\,. \end{aligned} \tag{A9}$$

A.1.1
Uncertainty Principle

Let us calculate the uncertainties of the momentum $\hat{p}$ and coordinate $\hat{x}$. By definition

$$\begin{aligned} (\Delta p)^2 &= \langle p^2\rangle - \langle p\rangle^2\,, \\ (\Delta x)^2 &= \langle x^2\rangle - \langle x\rangle^2\,, \end{aligned} \tag{A10}$$

where

$$\begin{aligned} \langle p\rangle &= \langle n|\frac{p_0}{i}(\hat{a} - \hat{a}^\dagger)|n\rangle = 0\,, \\ \langle p^2\rangle &= \langle n|p_0^2(\hat{a}\hat{a}^\dagger + \hat{a}^\dagger\hat{a})|n\rangle = \left(n + \frac{1}{2}\right)m\omega\hbar\,, \end{aligned} \tag{A11}$$

and

$$\begin{aligned} \langle x\rangle &= \langle n|x_0(\hat{a} + \hat{a}^\dagger)|n\rangle = 0\,, \\ \langle x^2\rangle &= \langle n|x_0^2(\hat{a}\hat{a}^\dagger + \hat{a}^\dagger\hat{a})|n\rangle = \left(n + \frac{1}{2}\right)\frac{\hbar}{m\omega}\,. \end{aligned} \tag{A12}$$

The uncertainty product is

$$\Delta p\Delta x = \left(n + \frac{1}{2}\right)\hbar\,. \tag{A13}$$

Therefore, the uncertainty grows linearly with the growth of n.

A.1.2
Coherent States and Uncertainties

Coherent states were introduced and developed in quantum optics (see, for example, [5, 176, 177]). The coherent state is defined by

$$\begin{aligned} a|\alpha\rangle &= \alpha|\alpha\rangle\,, \\ \langle\alpha|a^{\dagger} &= \langle\alpha|\alpha^{*}\,, \end{aligned} \tag{A14}$$

where α is an eigenvalue (a complex number) of the annihilation operator. We can represent the mean values of operators using the following relations:

$$\langle\alpha|a|\alpha\rangle = \alpha\,, \quad \langle\alpha|a^{\dagger}|\alpha\rangle = \alpha^{*}\,. \tag{A15}$$

In general, for ordered operators one has

$$\langle\alpha|(a^{\dagger})^{\nu_1}(a)^{\nu_2}|\alpha\rangle = (\alpha^{*})^{\nu_1}(\alpha)^{\nu_2}\,, \tag{A16}$$

where ν_1 and ν_2 are the natural numbers.

In the case of nonordered operators, for example, $aa^{\dagger}$ we obtain

$$\langle\alpha|\hat{a}\hat{a}^{\dagger}|\alpha\rangle = \langle\alpha|\hat{a}^{\dagger}\hat{a} + 1|\alpha\rangle = \alpha^{*}\alpha + 1\,. \tag{A17}$$

Let us calculate the uncertainties (A10) for a coherent state. For the momentum one has

$$\langle p\rangle = \langle\alpha|\frac{p_0}{i}(\hat{a} - \hat{a}^{\dagger})|\alpha\rangle = \frac{p_0}{i}(\alpha - \alpha^{*})$$

and

$$\begin{aligned} \langle p^2\rangle &= -\langle\alpha|p_0^2(\hat{a}\hat{a} - \hat{a}\hat{a}^{\dagger} - \hat{a}^{\dagger}\hat{a} + \hat{a}^{\dagger}\hat{a}^{\dagger})|\alpha\rangle \\ &= (2\alpha^{*}\alpha + 1 - \alpha\alpha - \alpha^{*}\alpha^{*})p_0^2\,, \end{aligned}$$

from which we obtain

$$\begin{aligned} (\Delta p)^2 &= \left[1 - (\alpha - \alpha^{*})^2\right] p_0^2 + (\alpha - \alpha^{*})^2 p_0^2 \\ &= p_0^2 = \frac{m\hbar\omega}{2}\,. \end{aligned} \tag{A18}$$

For the coordinate one has

$$\langle x\rangle = \langle\alpha|x_0(\hat{a} + \hat{a}^{\dagger})|\alpha\rangle = x_0(\alpha + \alpha^{*})$$

and

$$\begin{aligned} \langle x^2\rangle &= \langle\alpha|x_0^2(\hat{a}\hat{a} + \hat{a}\hat{a}^{\dagger} + \hat{a}^{\dagger}\hat{a} + \hat{a}^{\dagger}\hat{a}^{\dagger})|\alpha\rangle \\ &= \left(2\alpha^{*}\alpha + 1 + \alpha\alpha + \alpha^{*}\alpha^{*}\right) p_0^2\,, \end{aligned}$$

from which we obtain

$$(\Delta x)^2 = \left[1 + (\alpha + \alpha^*)^2\right] x_0^2 - (\alpha + \alpha^*)^2 x_0^2$$
$$= x_0^2 = \frac{\hbar}{2m\omega} . \tag{A19}$$

The uncertainty product is always equal to

$$\Delta p \Delta x = \frac{\hbar}{2} . \tag{A20}$$

Thus, any coherent state is the minimum-uncertainty state.

Substituting operators of creation and annihilation by corresponding complex numbers (eigenvalues of coherent states), we continue to work with the operators mean values in the coherent state representation. However, the information about an order of operation will be lost. This information is not important for the excited states of oscillator with $N \gg 1$.

It is easy to generalize the above conclusion for a system of harmonic oscillators, where, for example,

$$\hat{a}_1 \hat{a}_k |\alpha_1 \alpha_2 \dots \alpha_n\rangle = \alpha_1 \alpha_k |\alpha_1 \alpha_2 \dots \alpha_n\rangle ,$$
$$\langle \alpha_1 \alpha_2 \dots \alpha_n | \hat{a}_1^\dagger \hat{a}_k^\dagger = \langle \alpha_1 \alpha_2 \dots \alpha_n | \alpha_1^* \alpha_k^* , \tag{A21}$$

and

$$\langle \alpha_1 \alpha_2 \dots \alpha_n | \hat{a}_1^\dagger \hat{a}_k^\dagger \hat{a}_1 \hat{a}_k | \alpha_1 \alpha_2 \dots \alpha_n \rangle = \alpha_1^* \alpha_1 \alpha_k^* \alpha_k . \tag{A22}$$

Coherent states of oscillators is a powerful theoretical tool to analyze statistical properties of quantum system. In application to magnon systems, coherent states have been used in [178–180].

Appendix B
Dipolar Sums

In order to get an explicit expression for the spin-wave spectrum let us consider the following calculations:

$$\frac{V_s}{\mathcal{N}}\sum_{r}\frac{r^2-3z^2}{r^5} = -\frac{V_s}{\mathcal{N}}\sum_{r}\frac{\partial^2}{\partial z^2}\frac{1}{r}$$
$$= -\frac{V_s}{\mathcal{N}}\sum_{|r|<\rho}\frac{\partial^2}{\partial z^2}\frac{1}{r} - \int_{|r|>\rho}\left(\frac{\partial^2}{\partial z^2}\frac{1}{r}\right)d\boldsymbol{r}. \tag{B1}$$

Here $d_c \ll \rho \ll L$, L is the characteristic linear size of the magnetic moment change [9]. The second term in (B1) can be calculated in the following way

$$\int_{|r|>\rho}\left(\frac{\partial^2}{\partial z^2}\frac{1}{r}\right)d\boldsymbol{r} = \int_{|r|>\rho}\operatorname{div}\left(\frac{z\boldsymbol{e}_z}{r^3}\right)d\boldsymbol{r} = \oint_S \frac{z}{r^3}(\boldsymbol{e}_z d\boldsymbol{S})$$

and represented as

$$\oint_S \frac{z}{r^3}(\boldsymbol{e}_z d\boldsymbol{S}) = \oint_{\text{surface}} \frac{z}{r^3}(\boldsymbol{e}_z d\boldsymbol{S}) - \oint_{\text{sphere}} \frac{z}{r^3}(\boldsymbol{e}_z d\boldsymbol{S}).$$

Taking into account that

$$\oint_{\text{sphere}} \frac{z}{r^3}(\boldsymbol{e}_z d\boldsymbol{S}) = \frac{4\pi}{3}$$

and the following notations

$$\beta_{zz} \equiv \frac{V_s}{\mathcal{N}}\sum_{|r|<\rho}\frac{\partial^2}{\partial z^2}\frac{1}{r},$$

$$N_{zz} \equiv \frac{1}{4\pi}\oint_{\text{surface}} \frac{z}{r^3}(\boldsymbol{e}_z d\boldsymbol{S}),$$

we obtain

$$\frac{V_s}{\mathcal{N}}\sum_{r}\frac{r^2-3z^2}{r^5} = -\beta_{zz} + 4\pi\left(N_{zz}-\frac{1}{3}\right), \tag{B2}$$

Nonequilibrium Magnons, First Edition. Vladimir L. Safonov.

where V_s is the volume of the sample, β_{zz} the factor of local dipolar anisotropy and N_{zz} the demagnetizing factor of the sample. Analogous formulas can be obtained for xx and yy combinations:

$$\frac{V_s}{\mathcal{N}}\sum_{\boldsymbol{r}}\frac{r^2-3x^2}{r^5}=-\beta_{xx}+4\pi\left(N_{xx}-\frac{1}{3}\right), \tag{B3}$$

$$\frac{V_s}{\mathcal{N}}\sum_{\boldsymbol{r}}\frac{r^2-3y^2}{r^5}=-\beta_{yy}+4\pi\left(N_{yy}-\frac{1}{3}\right). \tag{B4}$$

All ellipsoids having principal axes in x, y, and z directions are characterized by N_{xx}, N_{yy} and N_{zz} with the following relation

$$N_{xx}+N_{yy}+N_{zz}=1\,. \tag{B5}$$

For a sphere one has $N_{xx}=N_{yy}=N_{zz}=1/3$, for a long (along z-axis) cylinder we have $N_{xx}=N_{yy}=1/2$, $N_{zz}=0$, and a thin (xy-plane) film is characterized by $N_{xx}=N_{yy}=0$, $N_{zz}=1$. For cubic symmetry $\beta_{ij}=0$.

Now let us consider the following calculation

$$\mathrm{I}_{zz}=\frac{V_s}{\mathcal{N}}\sum_{\boldsymbol{r}}\frac{r^2-3z^2}{r^5}e^{i\boldsymbol{k}\boldsymbol{r}}=-\beta_{zz}+\int\limits_{|\boldsymbol{r}|>\rho}\frac{(1-3\cos^2\theta)e^{i\boldsymbol{k}\boldsymbol{r}}}{r^3}d\boldsymbol{r}\,.$$

Here we can use the following formula

$$e^{i\boldsymbol{k}\boldsymbol{r}}=\sum_{n=0}^{\infty}(2n+1)i^n \times\sum_{m=-n}^{n}\frac{(n-|m|)!}{(n+|m|)!}\cos[m(\phi-\phi_k)]P_n^{|m|}(\cos\theta)P_n^{|m|}(\cos\theta_k)\mathrm{j}_n(kr)\,, \tag{B6}$$

where

$$\mathrm{j}_n(z)\equiv(-1)^n z^n\left(\frac{d}{zdz}\right)^n\left(\frac{\sin z}{z}\right) \tag{B7}$$

is the spherical Bessel function, and ϕ_k and θ_k are the spherical angles of $\boldsymbol{k}$.

Now we have

$$\mathrm{I}_{zz}=-\beta_{zz}-2\int\frac{d\phi\,d\cos\theta\,r^2dr}{r^3}P_2^0(\cos\theta)\sum_{n=0}^{\infty}(2n+1)i^n\,; \times\sum_{m=-n}^{n}\frac{(n-|m|)!}{(n+|m|)!}\cos[m(\phi-\phi_k)]P_n^{|m|}(\cos\theta)P_n^{|m|}(\cos\theta_k)\mathrm{j}_n(kr)\,. \tag{B8}$$

Taking into account that

$$\int\limits_{-1}^{+1}\left[P_2^0(x)\right]^2dx=\frac{2}{5}\,, \tag{B9}$$

one obtains

$$\mathrm{I}_{zz} = -\beta_{zz} + 2 \cdot 4\pi \cdot P_2^0(\cos\theta_k) \int \frac{dr}{r} \mathrm{j}_2(kr) \,. \tag{B10}$$

Then we can represent

$$\mathrm{j}_2(x) = x^2 \left(\frac{d}{x\,dx}\right)\left(\frac{\sin x}{x}\right) = -x\frac{d}{dx}\frac{\mathrm{j}_1(x)}{x} \tag{B11}$$

and obtain

$$\mathrm{I}_{zz} = -\beta_{zz} - 2 \cdot 4\pi \cdot P_2^0(\cos\theta_k) \left[\frac{\mathrm{j}_1(kr)}{kr}\right]_{k\rho}^{kR_s} \,, \tag{B12}$$

$$\mathrm{j}_1(x) = -\frac{x\cos x - \sin x}{x^2} = \begin{cases} 0 & \text{if} \quad x \to \infty \\ \frac{x}{3} & \text{if} \quad x \to 0 \end{cases} \,. \tag{B13}$$

Thus we get the following relation

$$\frac{V_s}{\mathcal{N}} \sum_r \frac{r^2 - 3z^2}{r^5} e^{i\mathbf{k}\mathbf{r}} = -\beta_{zz} + 4\pi \left[\left(\frac{k_z}{k}\right)^2 - \frac{1}{3}\right] \,. \tag{B14}$$

Using a similar calculation, one obtains

$$\frac{V_s}{\mathcal{N}} \sum_r \frac{r^2 - 3x^2}{r^5} e^{i\mathbf{k}\mathbf{r}} = -\beta_{xx} + 4\pi \left[\left(\frac{k_x}{k}\right)^2 - \frac{1}{3}\right] \tag{B15}$$

and

$$\frac{V_s}{\mathcal{N}} \sum_r \frac{r^2 - 3z^2}{r^5} e^{i\mathbf{k}\mathbf{r}} = -\beta_{zz} + 4\pi \left[\left(\frac{k_z}{k}\right)^2 - \frac{1}{3}\right] \,. \tag{B16}$$

For the case of mixed components we have

$$\frac{V_s}{\mathcal{N}} \sum_r \frac{3xy}{r^5} e^{i\mathbf{k}\mathbf{r}} = -4\pi \frac{k_x k_y}{k^2} \,, \tag{B17}$$

$$\frac{V_s}{\mathcal{N}} \sum_r \frac{3xz}{r^5} e^{i\mathbf{k}\mathbf{r}} = -4\pi \frac{k_x k_z}{k^2} \,, \tag{B18}$$

$$\frac{V_s}{\mathcal{N}} \sum_r \frac{3yz}{r^5} e^{i\mathbf{k}\mathbf{r}} = -4\pi \frac{k_y k_z}{k^2} \,. \tag{B19}$$

Appendix C
Unitary Transformations in Weakly Nonideal Bose Gases

Unitary transformations play an important role in solid state physics [181]. Using unitary transformation, it is possible to simplify a problem of interacting quasi particles by eliminating "inconvenient" terms from the initial Hamiltonian and constructing terms much more convenient for theoretical analysis.

A general form of unitary transformation can be written as

$$\tilde{\mathcal{H}}(\theta) = e^{\theta\mathcal{R}}\mathcal{H}e^{-\theta\mathcal{R}} , \tag{C1}$$

where $\mathcal{R}$, is an antihermitian operator ($\mathcal{R}^{\dagger} = -\mathcal{R}$) and θ a formal parameter.

This expression is the solution of the equation

$$\frac{d}{d\theta}\tilde{\mathcal{H}}(\theta) = [\mathcal{R}, \tilde{\mathcal{H}}(\theta)] \tag{C2}$$

with the initial condition $\tilde{\mathcal{H}}(0) = \mathcal{H}$.

One can write the most general form of $\tilde{\mathcal{H}}(\theta)$ as the expansion in terms of hermitian operator combinations with unknown θ-dependent coefficients and the most general form of $\mathcal{R}$ in terms of antihermitian operator combinations with free parameters.

Substituting these general expressions into (C2), one can obtain a set of linear differential equations for unknown coefficients in analogous operator compositions. Solving these equations with the initial conditions, we get the transformed Hamiltonian $\tilde{\mathcal{H}}(\theta)$. In order to eliminate inconvenient terms one needs to put their coefficients (for example, for $\theta = 1$) equal to zero. This condition defines the choice of $\mathcal{R}$.

The above procedure was first proposed for the spin Hamiltonian diagonalization [182–184] and was successfully used in the physics of magnetic excitations [185, 186], in the theory of superconductivity [187] and for eliminating three-boson interactions [188].

One can easily check that the quadratic Hamiltonian of the Bose system in $\boldsymbol{k}$-space

$$\mathcal{H}^{(2)} = \sum_{k}\left[A_k b_k^{\dagger} b_k + \frac{1}{2}\left(B_k b_k^{\dagger} b_{-k}^{\dagger} + B_k^{*} b_k b_{-k}\right)\right] \tag{C3}$$

is transformed to a diagonal form

$$\mathcal{H}_0 = \sum_k \varepsilon_k b_k^\dagger b_k \,, \tag{C4}$$

by the unitary transformation $U\mathcal{H}^{(2)}U^{-1}$, where

$$U = \exp\left[\sum_q \left(R_q b_q^\dagger b_{-q}^\dagger - R_q^* b_q b_{-q}\right)\right] ,$$

$$\varepsilon_k = \operatorname{sign}(A_k)\sqrt{A_k^2 - |B_k|^2} \,,$$

and

$$\tanh(4|R_k|) = -\frac{|B_k|}{A_k} \,,$$

$$\arg(R_k) = \arg(B_k) \,.$$

C.1
One-Component Bose Gas

Typical inconvenient terms in the Hamiltonians of the theory of magnons and magnetoelastic excitations in magneto-ordered crystals describe three-boson processes, which are forbidden by the laws of conservation of energy and momentum. Consider, for example, the Hamiltonian

$$\mathcal{H} = \mathcal{H}_0 + \mathcal{H}_1^{(3)} \,, \tag{C5}$$

$$\mathcal{H}_1^{(3)} = \frac{1}{3}\sum_{1,2,3}\left[\Psi_1(1,2,3) b_1^\dagger b_2^\dagger b_3^\dagger + \text{h.c.}\right]\Delta(1+2+3) \,, \tag{C6}$$

where $\Delta(\boldsymbol{k})$ is the Kronecker delta function which describes here the law of conservation of momentum $\mathbf{1}+\mathbf{2}+\mathbf{3} \equiv \boldsymbol{k}_1+\boldsymbol{k}_2+\boldsymbol{k}_3 = 0$. Three creating (and annihilating) boson terms $b_1^\dagger b_2^\dagger b_3^\dagger \equiv b_{k_1}^\dagger b_{k_2}^\dagger b_{k_3}^\dagger$ appear in the theory of a ferromagnet from the dipole–dipole interactions. For the energy spectrum $\varepsilon_k > 0$ the interaction terms in (C6) play no role in the magnon dynamics in the first approximation as far as

$$\varepsilon_1 + \varepsilon_2 + \varepsilon_3 \neq 0$$

for any

$$\boldsymbol{k}_1 + \boldsymbol{k}_2 + \boldsymbol{k}_3 = 0 \,.$$

It looks like the Hamiltonian $\mathcal{H}_1^{(3)}$ describes completely forbidden magnon interactions and plays no role in the magnon dynamics. Therefore it can be omitted in the analysis. However, this is the wrong impression. Applying a unitary transformation, we can exclude "ineffective" three-boson terms and construct effective four-boson interactions.

Let us write the general form of the Hamiltonian (C5) as follows

$$\tilde{\mathcal{H}}(\theta) = \sum_k \varepsilon_k b_k^\dagger b_k + \frac{1}{3}\sum_{1,2,3}\left[\tilde{\Psi}_1(1,2,3;\theta) b_1^\dagger b_2^\dagger b_3^\dagger + \text{h.c.}\right]\Delta(1+2+3)$$
$$+ \frac{1}{2}\sum_{1,2,3,4}\tilde{\Phi}(1,2,3,4;\theta) b_1^\dagger b_2^\dagger b_3 b_4 \Delta(1+2-3-4)\,. \tag{C7}$$

Here and below we omit all terms containing five and more Bose operators.

Let us make some general remarks.

a) As far as we consider a weakly interacting Bose gas, the amplitudes of interactions have to decrease with increasing number of Bose operators in the Hamiltonian. For example, for (C7) it looks like $\varepsilon_k \gg \tilde{\Psi}_1/3 \gg \tilde{\Phi}/2$. In this case the order of the creation and annihilation operators in the interaction terms is not so important. If we change the order of operators, say

$$b_1^\dagger b_3 b_2^\dagger b_4 = b_1^\dagger b_2^\dagger b_3 b_4 + b_1^\dagger b_4 \Delta(2-3)\,,$$

then the role of additional terms will be negligibly small relative to the "bare" quadratic terms.

b) It is easy to show that a result of commutation of combinations of Bose operators

$$\left[\underbrace{b^\dagger \cdots b^\dagger}_{n_1}\underbrace{b \ldots b}_{m_1},\quad \underbrace{b^\dagger \ldots b^\dagger}_{n_2}\underbrace{b \ldots b}_{m_2}\right]$$

consists of a sum of compositions containing $n_1 + m_1 + n_2 + m_2 - 2$ Bose operators: each of $n_1 + n_2 - 1$ the creation and $m_1 + m_2 - 1$ the annihilation operators. For example, for $n_2 = m_2 = 1$, the result of commutation contains compositions with the same numbers of creation (n_1) and annihilation (m_2) operators. This means that if we want to remove a term

$$\chi\,\underbrace{b^\dagger \ldots b^\dagger}_{n_1}\underbrace{b \ldots b}_{m_1} + \text{h.c.}$$

from the initial Hamiltonian, we should use the antihermitian operator of the form

$$R_\chi\,\underbrace{b^\dagger \ldots b^\dagger}_{n_1}\underbrace{b \ldots b}_{m_1} - \text{h.c.}$$

C.1.1
Three-Boson Annihilation

Thus in order to remove the ineffective three-boson term from (C7) we shall use the unitary transformation with the antihermitian operator

$$\mathcal{R}_1 = \sum_{1,2,3} \left[R_1(1,2,3) b_1^\dagger b_2^\dagger b_3^\dagger - \text{h.c.} \right] \Delta(1+2+3) \,. \tag{C8}$$

Calculating the following commutators

$$\left[R_1(q_1, q_2, q_3) b_{q_1}^\dagger b_{q_2}^\dagger b_{q_3}^\dagger - \text{h.c.}\,, \quad \varepsilon_k b_k^\dagger b_k \right]$$

and

$$\left[R_1(q_1, q_2, q_3) b_{q_1}^\dagger b_{q_2}^\dagger b_{q_3}^\dagger - \text{h.c.}\,, \quad \tilde{\Psi}_1(k_1, k_2, k_3; \theta) b_{k_1}^\dagger b_{k_2}^\dagger b_{k_3}^\dagger + \text{h.c.} \right] ,$$

from (C2) we can obtain the equations

$$\begin{aligned} \frac{d}{d\theta} \tilde{\Psi}_1(1,2,3;\theta) &= -3(\varepsilon_1 + \varepsilon_2 + \varepsilon_3) R_1(1,2,3) \,, \\ \frac{d}{d\theta} \tilde{\Phi}(1,2,3,4;\theta) &= -6 R_1(1,2,-1-2) \tilde{\Psi}_1^*(3,4,-3-4;\theta) \\ &\quad -6 R_1^*(3,4,-3-4) \tilde{\Psi}_1(1,2,-1-2;\theta) \,. \end{aligned} \tag{C9}$$

Solving these equations with the initial conditions

$$\tilde{\Psi}_1(\theta = 0) = \Psi_1 \,, \quad \text{and} \quad \tilde{\Phi}(\theta = 0) = 0 \,,$$

and demanding $\tilde{\Psi}_1(\theta = 1) = 0$, we get

$$R_1(1,2,3) = \frac{\Psi_1(1,2,3)}{3(\varepsilon_1 + \varepsilon_2 + \varepsilon_3)} \tag{C10}$$

and

$$\begin{aligned} \tilde{\Phi}(1,2,3,4;1) = &- \Psi_1(1,2,-1-2) \Psi_1^*(3,4,-3-4) \\ &\times \left(\frac{1}{\varepsilon_1 + \varepsilon_2 + \varepsilon_{-1-2}} + \frac{1}{\varepsilon_3 + \varepsilon_4 + \varepsilon_{-3-4}} \right) . \end{aligned} \tag{C11}$$

If the initial Hamiltonian (C6) contains a "bare" four-boson interaction

$$\mathcal{H}_1^{(4)} = \frac{1}{2} \sum_{1,2,3,4} \Phi(1,2,3,4) b_1^\dagger b_2^\dagger b_3 b_4 \Delta(1+2-3-4) \tag{C12}$$

then the effective amplitude of boson–boson scattering in the transformed $\tilde{\mathcal{H}}_1^{(4)}$ $(\theta = 1)$, is

$$\Phi(1,2,3,4) + \tilde{\Phi}(1,2,3,4;1) \,. \tag{C13}$$

C.1.2
The Confluence and Decay Processes

Let us now consider the three-boson interactions energy of the form,

$$\mathcal{H}_2^{(3)} = \sum_{1,2,3} \left[\Psi_2(1,2,3) b_1^\dagger b_2^\dagger b_3 + \text{h.c.} \right] \Delta(1+2-3) . \tag{C14}$$

This Hamiltonian contains terms which describe the three-boson process

$$\varepsilon_1 + \varepsilon_2 - \varepsilon_3 = 0 ,$$
$$\boldsymbol{k}_1 + \boldsymbol{k}_2 - \boldsymbol{k}_3 = 0 .$$

However, the Hamiltonian (C14) contains also forbidden terms for which

$$\Delta\varepsilon(\boldsymbol{k}_1, \boldsymbol{k}_2, \boldsymbol{k}_3) \equiv \varepsilon_1 + \varepsilon_2 - \varepsilon_3 \neq 0 \quad \text{for} \quad \boldsymbol{k}_1 + \boldsymbol{k}_2 - \boldsymbol{k}_3 = 0 .$$

In order to remove these forbidden three-boson processes, we shall use the unitary transformation with the antihermitian operator

$$\mathcal{R}_2 = \sum_{1,2,3} \left[R_2(1,2,3) b_1^\dagger b_2^\dagger b_3 - \text{h.c.} \right] \Delta(1+2-3) . \tag{C15}$$

We introduce a general form of Hamiltonian containing $\mathcal{H}_2^{(3)}$ with $\tilde{\Psi}_2(1,2,3;\theta)$, and analogous to the above calculations, one obtains

$$R_2(1,2,3) = \frac{\Psi_2(1,2,3)}{\varepsilon_1 + \varepsilon_2 - \varepsilon_3} . \tag{C16}$$

A corresponding additional part to the effective four-boson amplitude (C13) has the form:

$$\begin{aligned}
&\Psi_2(1,2,1+2)\Psi_2^*(3,4,3+4) \\
&\quad \times \left(\frac{1}{\varepsilon_1 + \varepsilon_2 - \varepsilon_{1+2}} + \frac{1}{\varepsilon_3 + \varepsilon_4 - \varepsilon_{3+4}} \right) \\
&- \Psi_2(1,3-1,3)\Psi_2^*(4,2-4,2) \\
&\quad \times \left(\frac{1}{\varepsilon_1 + \varepsilon_{3-1} - \varepsilon_3} + \frac{1}{\varepsilon_4 + \varepsilon_{2-4} - \varepsilon_2} \right) \\
&- \Psi_2(1,4-1,4)\Psi_2^*(3,2-3,2) \\
&\quad \times \left(\frac{1}{\varepsilon_1 + \varepsilon_{4-1} - \varepsilon_4} + \frac{1}{\varepsilon_3 + \varepsilon_{2-3} - \varepsilon_2} \right) \\
&- \Psi_2(2,3-2,3)\Psi_2^*(4,1-4,1) \\
&\quad \times \left(\frac{1}{\varepsilon_2 + \varepsilon_{3-2} - \varepsilon_3} + \frac{1}{\varepsilon_4 + \varepsilon_{1-4} - \varepsilon_1} \right) \\
&- \Psi_2(2,4-2,4)\Psi_2^*(3,1-3,1) \\
&\quad \times \left(\frac{1}{\varepsilon_2 + \varepsilon_{4-2} - \varepsilon_4} + \frac{1}{\varepsilon_3 + \varepsilon_{1-3} - \varepsilon_1} \right) .
\end{aligned} \tag{C17}$$

This expression is valid if $\varepsilon_k \gg \Psi\Psi^*/\Delta\varepsilon$.

C.2
Two-Component Bose Gas

Now we turn our attention to a two-component Bose gas with the Hamiltonian

$$\mathcal{H} = \sum_k (\varepsilon_k b_k^\dagger b_k + e_k c_k^\dagger c_k)$$
$$+ \sum_{1,2,3} \left[\Psi_3(1,2,3) b_1^\dagger b_2^\dagger c_3^\dagger + \text{h.c.} \right] \Delta(1+2+3)$$
$$+ \sum_{1,2,3} \left[\Psi_4(1,2,3) b_1^\dagger b_2^\dagger c_3 + \text{h.c.} \right] \Delta(1+2-3)$$
$$+ 2 \sum_{1,2,3} \left[\Psi_5(1,2,3) b_1^\dagger b_2 c_3^\dagger + \text{h.c.} \right] \Delta(1-2+3) , \quad \text{(C18)}$$

where e_k is the energy and $c_k^\dagger$, c_k are the corresponding Bose operators which commute with $b_q^\dagger$, b_q. We are interested in eliminating the forbidden three-boson terms from (C18) and obtaining additional parts to four-boson amplitudes in the Hamiltonian (C13) and

$$\mathcal{H}_2^{(4)} = \sum_{1,2,3,4} \Upsilon(1,2,3,4) b_1^\dagger c_2^\dagger b_3 c_4 \Delta(1+2-3-4) . \quad \text{(C19)}$$

It is convenient to remove the ineffective three-boson terms from (C18), by means of several unitary transformations. The antihermitian operators can be taken in the forms

$$\mathcal{R}_3 = \sum_{1,2,3} \left[R_3(1,2,3) b_1^\dagger b_2^\dagger c_3^\dagger - \text{h.c.} \right] \Delta(1+2+3) ,$$
$$\mathcal{R}_4 = \sum_{1,2,3} \left[R_4(1,2,3) b_1^\dagger b_2^\dagger c_3 - \text{h.c.} \right] \Delta(1+2-3) ,$$
$$\mathcal{R}_5 = \sum_{1,2,3} \left[R_5(1,2,3) b_1^\dagger b_2 c_3^\dagger - \text{h.c.} \right] \Delta(1-2+3) . \quad \text{(C20)}$$

Analogous to the above calculations, we obtain

$$R_3(1,2,3) = \frac{\Psi_3(1,2,3)}{\varepsilon_1 + \varepsilon_2 + e_3} ,$$
$$R_4(1,2,3) = \frac{\Psi_4(1,2,3)}{\varepsilon_1 + \varepsilon_2 - e_3} ,$$
$$R_5(1,2,3) = \frac{2\Psi_5(1,2,3)}{\varepsilon_1 - \varepsilon_2 + e_3} . \quad \text{(C21)}$$

Corresponding additional parts to $\Phi(1,2,3,4)$ in (C13) can be written as

$$
\begin{aligned}
&-\Psi_3(1,2,-1,-2)\,\Psi_3^*(3,4,-3-4)\\
&\quad\times\left(\frac{1}{\varepsilon_1+\varepsilon_2+e_{-1-2}}+\frac{1}{\varepsilon_3+\varepsilon_4+e_{-3-4}}\right)\\
&+\Psi_4(1,2,1+2)\,\Psi_4^*(3,4,3+4)\\
&\quad\times\left(\frac{1}{\varepsilon_1+\varepsilon_2-e_{1+2}}+\frac{1}{\varepsilon_3+\varepsilon_4-e_{3+4}}\right)\\
&-\Psi_5(1,3,3-1)\,\Psi_5^*(4,2,2-4)\\
&\quad\times\left(\frac{1}{\varepsilon_1-\varepsilon_3+e_{3-1}}+\frac{1}{\varepsilon_4-\varepsilon_2+e_{2-4}}\right)\\
&-\Psi_5(1,4,4-1)\,\Psi_5^*(3,2,2-3)\\
&\quad\times\left(\frac{1}{\varepsilon_1-\varepsilon_4+e_{4-1}}+\frac{1}{\varepsilon_3-\varepsilon_2+e_{2-3}}\right)\\
&-\Psi_5(2,3,3-2)\,\Psi_5^*(4,1,1-4)\\
&\quad\times\left(\frac{1}{\varepsilon_2-\varepsilon_3+e_{3-2}}+\frac{1}{\varepsilon_4-\varepsilon_1+e_{1-4}}\right)\\
&-\Psi_5(2,4,4-2)\,\Psi_5^*(3,1,1-3)\\
&\quad\times\left(\frac{1}{\varepsilon_2-\varepsilon_4+e_{4-2}}+\frac{1}{\varepsilon_3-\varepsilon_1+e_{1-3}}\right)\,.
\end{aligned}
\tag{C22}
$$

For the additional parts to $\Upsilon(1,2,3,4)$ in (C19) we get

$$
\begin{aligned}
&-2\Psi_3(\boldsymbol{k}_1,-\boldsymbol{k}_1-\boldsymbol{k}_2,\boldsymbol{k}_2)\,\Psi_3^*(\boldsymbol{k}_3,-\boldsymbol{k}_3-\boldsymbol{k}_4,\boldsymbol{k}_4)\\
&\quad\times\left(\frac{1}{\varepsilon_{k_1}+\varepsilon_{-k1-k_2}+e_{k_2}}+\frac{1}{\varepsilon_{k_3}+\varepsilon_{-k_3-k_4}+e_{k_4}}\right)\\
&-2\Psi_4(\boldsymbol{k}_1,\boldsymbol{k}_4-\boldsymbol{k}_1,\boldsymbol{k}_4)\,\Psi_4^*(\boldsymbol{k}_3,\boldsymbol{k}_2-\boldsymbol{k}_3,\boldsymbol{k}_2)\\
&\quad\times\left(\frac{1}{\varepsilon_{k_1}+\varepsilon_{k_4-k_1}-e_{k_4}}+\frac{1}{\varepsilon_{k_3}+\varepsilon_{k_2-k_3}-e_{k_2}}\right)\\
&+2\Psi_5(1,1+2,2)\,\Psi_5^*(3,3+4,4)\\
&\quad\times\left(\frac{1}{\varepsilon_1-\varepsilon_{1+2}+e_2}+\frac{1}{\varepsilon_3-\varepsilon_{3+4}+e_4}\right)\\
&\quad-2\Psi_5(3-2,3,2)\,\Psi_5^*(1-4,1,4)\\
&\times\left(\frac{1}{\varepsilon_{k_3-k_2}-\varepsilon_{k_3}+e_{k_2}}+\frac{1}{\varepsilon_{k_1-k_4}-\varepsilon_{k_1}+e_{k_2}}\right)\,.
\end{aligned}
\tag{C23}
$$

C.3
Concluding Remarks

Thus we have demonstrated how to eliminate the three-boson forbidden processes from a Hamiltonian. The procedure was based on the solution of a system of linear differential equations. This made it possible to take into account all terms in the expansion of the unitary operator.

A modification to the above method of unitary transformation (C1), the so-called flow equation method, was proposed in [192]. The difference is a parameter dependence of the generator $R(\theta)$ on the formal parameter θ in (C2). A detailed analysis [193] has shown that the modified method in many cases gives the same results, but with more cumbersome calculations.

Classical canonical transformations for eliminating nonresonant terms have been considered in [4, 72, 73]. Krasitskii [189, 190] has developed a general method of canonical transformation for the classical weakly nonlinear wave system. In order to remove the forbidden three-wave interaction terms from the Hamiltonian he considered the expansion for complex wave amplitudes

$$a_k = b_k + I^{(2)}(\boldsymbol{k}) + I^{(3)}(\boldsymbol{k}) \,, \tag{C24}$$

where $I^{(n)}(\boldsymbol{k})$ contains terms with different compositions of n new complex amplitudes. The transformation is canonical if certain conditions for the coefficients of these compositions in the expansion (C24) are satisfied.

There is another way to construct a nonlinear canonical transformation for a classical system. According to Bohm's general theory of collective coordinates [191], we can express old coordinates of the system in terms of new variables

$$q_{\text{old}} = e^{-\mathcal{R}} q_{\text{new}} e^{\mathcal{R}} = \left\{ q + [q, \mathcal{R}] + \frac{1}{2}[(q, \mathcal{R}), \mathcal{R}] + \cdots \right\}_{\text{new}} .$$

Here we can use Poisson brackets multiplied by $-i\hbar$ instead of the commutator.

Our results for the additional part to $\Phi(\boldsymbol{k}_1, \boldsymbol{k}_2, \boldsymbol{k}_3, \boldsymbol{k}_4)$ coincide with those in [189] if we use the additional conditions: $\Psi_1 = \Psi_1^*$, $\Psi_2 = \Psi_2^*$ and $\varepsilon_{\boldsymbol{k}_1} + \varepsilon_{\boldsymbol{k}_2} = \varepsilon_{\boldsymbol{k}_3} + \varepsilon_{\boldsymbol{k}_4}$, for $\boldsymbol{k}_1 + \boldsymbol{k}_2 = \boldsymbol{k}_3 + \boldsymbol{k}_4$.

Appendix D
Magnetization Dynamic Equation

In this appendix we derive a dynamic magnetization equation that describes large magnetization motions including magnetization reversal. We assume that the total magnetization $|\boldsymbol{M}|$ is a constant during reversal. All spins perform a coherent motion and the role of nonuniform spin motions is neglected. An excess of magnetic energy goes directly to a nonmagnetic thermal bath.

A conventional theoretical tool to study magnetization motion is based on the phenomenological Landau–Lifshitz equation [14] or, its equivalent modification with the Gilbert form of relaxation [62, 63] (so-called LLG equation). These equation conserve the absolute value of magnetization ($|\boldsymbol{M}| = \text{const}$) in a single-domain region. Both equations were introduced (1) for small magnetization motions and (2) for the case of uniaxial magnetic symmetry. The energy losses are defined by an isotropic phenomenological damping fitting parameter α (so-called damping constant).

A theoretical approach [64, 146, 194] has been developed to correct the limitations of the Landau–Lifshitz–Gilbert theory. The main idea was to represent the magnetization dynamics as the motion of a damped nonlinear oscillator. The oscillator model is a convenient tool to establish a "bridge" between the microscopic physics, where the rotational oscillator complex variables naturally describe spin excitations and the macroscopic magnetization dynamics.

Let us consider small-amplitude magnetization motions of a single-domain ferromagnetic particle in the vicinity of equilibrium state $\boldsymbol{M}||\hat{z}$, where $\hat{z}$ is the unit vector in the equilibrium direction. The magnetization rotation around the effective field in this case, in general, is elliptical and the magnetic energy $\mathcal{E}$ can be represented as a quadratic form:

$$\frac{\mathcal{E}}{V} = \frac{H_x}{2M_\text{s}} M_x^2 + \frac{H_y}{2M_\text{s}} M_y^2 \,. \tag{D1}$$

Here $\boldsymbol{M}$ is the magnetization vector, M_x and M_y are the components of this vector in the plane perpendicular to the equilibrium direction, M_s is the saturation magnetization and V is the particle volume. H_x and H_y are positive Kittel "stiffness" fields, which include both microscopic and shape anisotropies and the external magnetic field.

Nonequilibrium Magnons, First Edition. Vladimir L. Safonov.

We can write the Holstein–Primakoff transformation for $\boldsymbol{S} = \boldsymbol{M}V/\hbar\gamma$ in the form:

$$
\begin{aligned}
S^+ &= a\sqrt{S + S_z}\,, \\
S^- &= a^*\sqrt{S + S_z}\,, \\
S_z &= S - a^*a\,,
\end{aligned}
\tag{D2}
$$

where $S^\pm = S_x \pm i S_y$, and a^* and a are the complex amplitudes that describe uniform precession of $\boldsymbol{S}$.

The magnetic energy (D1) can be written in the quadratic form:

$$
\frac{\mathcal{E}}{\hbar} = \mathcal{A}a^*a + \frac{\mathcal{B}}{2}(aa + a^*a^*)\,,
\tag{D3}
$$

where

$$
\mathcal{A} = \frac{\gamma(H_x + H_y)}{2}\,,
\qquad
\mathcal{B} = \frac{\gamma(H_x - H_y)}{2}\,.
$$

The nondiagonal terms in (D3) are eliminated by the linear canonical transformation:

$$
a = uc + vc^*\,, \quad a^* = uc^* + vc\,,
\tag{D4}
$$

where

$$
u = \sqrt{\frac{\mathcal{A} + \omega_0}{2\omega_0}}\,, \quad v = -\frac{\mathcal{B}}{|\mathcal{B}|}\sqrt{\frac{\mathcal{A} - \omega_0}{2\omega_0}}\,.
$$

The energy in terms of the normal mode coordinates c and c^* is simply:

$$
\mathcal{E} = \hbar\omega_0 c^* c\,,
\tag{D5}
$$

where

$$
\omega_0 = \sqrt{\mathcal{A}^2 - \mathcal{B}^2} = \gamma\sqrt{H_x H_y}
$$

is the ferromagnetic resonance frequency.

The dynamic equations for c and c^* are independent and can be written as:

$$
\begin{aligned}
\frac{dc}{dt} + \eta c &= -i\omega_0 c\,, \\
\frac{dc^*}{dt} + \eta c^* &= i\omega_0 c^*\,.
\end{aligned}
\tag{D6}
$$

Here η is the linear relaxation rate, which can be found microscopically (Chapter 3).

In the case of large magnetization motion we can write a corresponding nonlinear oscillator equation in the form:

$$\frac{dc}{dt} + \eta(N)c = G(c, c^*) \,, \tag{D7}$$

where $N \equiv c^* c$ and $G(c, c^*)$ corresponds to the gyromagnetic term $-\gamma \boldsymbol{M} \times \boldsymbol{H}_{\text{eff}}$. The nonlinear relaxation rate $\eta(N)$ can be estimated from the known relaxation process for the uniform precession [194].

Using back transformations (D4) and (D2), we can derive an equation (corresponding to (D6)) in terms of unit magnetization vector ($\boldsymbol{m} = \boldsymbol{M}/M_s = \boldsymbol{S}/S$) components [152]:

$$\frac{d\boldsymbol{m}}{dt} = -\gamma \boldsymbol{m} \times \boldsymbol{H}_{\text{eff}} - \overleftrightarrow{\Gamma} \cdot (\boldsymbol{m} - \hat{z}_0) \,, \tag{D8}$$

where

$$\overleftrightarrow{\Gamma} = 2\eta(N) \begin{pmatrix} \zeta & 0 & 0 \\ 0 & \zeta & 0 \\ 0 & 0 & 1 \end{pmatrix} , \quad \zeta = \frac{m_z}{1 + m_z} \tag{D9}$$

and

$$N = \frac{\mathcal{A}}{\omega_0}(1 - m_z) + \frac{\mathcal{B}}{\omega_0} \frac{m_x^2 - m_y^2}{1 + m_z} \,. \tag{D10}$$

Note that (D8) conserves the magnitude of $\boldsymbol{m}$. For small deviations from equilibrium, when $m_z \simeq 1$, this equations exactly correspond to Bloch–Bloembergen equations with $\eta(0) = 1/T_2$ and $1/T_1 = 2/T_2$.

From (D8) we see that one can expect the anomalously large damping in the case of 180° reversal when $1 + m_z \to 0$ (see, also [194]).

It should be emphasized that the microscopic damping mechanism imposes a specific term in the magnetization dynamic equation. The case of impurity relaxation mechanism for dynamic magnetization reversal in a single-domain grain was studied in [195].

One more important question that should be mentioned here is the role of fluctuation-dissipation relations for the magnetization dynamics with Landau—Lifshitz–Gilbert and Bloch–Bloembergen damping terms. Usually these relations are very important for studying the stochastic properties of magnetic media using micromagnetic methods. As demonstrated in [196], the use of the Callen–Welton fluctuation-dissipation theorem [197], which was proven for Hamiltonian systems only, can give an inconsistent result for magnetic systems with dissipation.

Appendix E
A Parametric Pair Single-Mode Realization

The Lie algebra of $SU(1,1)$ group is used in many branches of physics (see, for example, [11, 198–200]). This algebra consists of the three hyperbolic operators K_0, K_+, and K_-, which satisfy the following commutation relations:

$$[K_0, K_\pm] = \pm K_\pm \, ,$$
$$[K_+, K_-] = -2K_0 \, . \tag{E1}$$

The Casimir invariant for this algebra is

$$C = K_0^2 - \frac{1}{2}(K_+ K_- + K_- K_+) \, . \tag{E2}$$

The two-mode representation of this algebra exactly corresponds to the parametric pair operators:

$$K_- = b_k b_{-k} \, ,$$
$$K_+ = b_k^\dagger b_{-k}^\dagger \, ,$$
$$K_0 = \frac{1}{2}\left(b_k^\dagger b_k + b_{-k}^\dagger b_{-k} + 1\right) \tag{E3}$$

with

$$C(\boldsymbol{k}) = -\frac{1}{4} + \frac{\left(b_k^\dagger b_k - b_{-k}^\dagger b_{-k}\right)^2}{4} \, . \tag{E4}$$

The relations (E1) and (E2) resemble the spin operator commutations

$$[S_z, S_\pm] = \pm S_\pm \, ,$$
$$[S_+, S_-] = 2S_z \, , \tag{E5}$$

and the following invariant

$$S(S+1) = S_z^2 + \frac{1}{2}(S_+ S_- + S_- S_+) \, . \tag{E6}$$

Nonequilibrium Magnons, First Edition. Vladimir L. Safonov.

The $SU(1,1)$ algebra has a single-bosonic realization [199]:

$$\begin{aligned} K_- &= (2k + a^\dagger a)^{1/2} a\,, \\ K_+ &= a^\dagger (2k + a^\dagger a)^{1/2}\,, \\ K_0 &= k + a^\dagger a\,, \end{aligned} \qquad \text{(E7)}$$

$$C = k(k-1)\,, \quad k = \frac{1}{2}\left(l + \frac{N}{2}\right) \quad l = 0, 1, 2, \ldots \qquad \text{(E8)}$$

Here $a^\dagger$ and a are the creation and annihilation Bose operators, respectively. This representation is an analog of the Holstein–Primakoff representation for spin operators:

$$\begin{aligned} S_- &= (2S - a^+ a)^{1/2} a\,, \\ S_+ &= a^+ (2S + a^+ a)^{1/2}\,, \\ S_z &= -S + a^+ a\,. \end{aligned} \qquad \text{(E9)}$$

E.1 A Single-Mode Representation

There is another important single-mode representation for spin proposed by Villain [154]. Here we consider an analog of this single-mode representation for the hyperbolic operators and consider its application.

If X is a coordinate and $P = -i d/dX$ is the momentum operator, the Villain representation can be written as

$$\begin{aligned} S_- &= \sqrt{\left(S + \frac{1}{2}\right)^2 - \left(P - \frac{1}{2}\right)^2} \exp(-iX)\,, \\ S_+ &= \exp(iX) \sqrt{\left(S + \frac{1}{2}\right)^2 - \left(P - \frac{1}{2}\right)^2}\,, \\ S_z &= P\,. \end{aligned} \qquad \text{(E10)}$$

Due to the similarity between spin and hyperbolic operators mentioned above, one can construct the following single-mode representation:

$$\begin{aligned} K_- &= (P + P_0) \exp(-iX)\,, \\ K_+ &= \exp(iX)(P + P_0^*)\,, \\ K_0 &= P + \frac{(P_0 + P_0^*)}{2} - \frac{1}{2} \end{aligned} \qquad \text{(E11)}$$

with

$$C = -\frac{1}{4} + \frac{(P_0 - P_0^*)^2}{4}\,. \qquad \text{(E12)}$$

Here P_0 is a complex number. Taking into account that $[X, P] = i$ and

$$\exp(i\beta X) P^n \exp(-i\beta X) = (P - \beta)^n , \tag{E13}$$

where β is a formal parameter and $n = 1, 2, \ldots$, it is simple to check out the validity of commutations (E1) for (E11).

Another form of representation (E11) may be obtained if we change $X \to -P$ and $P \to X$.

Note that Perelomov [198] mentioned a single-mode realization of $SU(1,1)$ in the form:

$$\begin{aligned} K_0 &= -\frac{id}{d\theta} , \\ K_\pm &= -i \exp(\pm i\theta) \frac{d}{d\theta} \mp \left(-\frac{1}{2} + i\lambda\right) \exp(\pm i\theta) \end{aligned} \tag{E14}$$

with

$$C = -\frac{1}{4} - \frac{\lambda^2}{4} , \quad \lambda > 0 . \tag{E15}$$

After simple algebra, one can find that (E14) can be transformed to the form (E11) with $\theta = X$ and $P_0 = 1/2 + i\lambda$.

Substituting

$$\begin{aligned} Q &= \frac{a + a^\dagger}{\sqrt{2}} , \\ P &= \frac{a - a^\dagger}{i\sqrt{2}} , \end{aligned} \tag{E16}$$

we obtain the following representations of the hyperbolic operators in terms of Bose operators a and $a^\dagger$:

$$\begin{aligned} K_- &= \left(\frac{a - a^\dagger}{i\sqrt{2}} + P_0\right) \exp\left(-i\frac{a + a^\dagger}{\sqrt{2}}\right) , \\ K_+ &= \exp\left(i\frac{a + a^\dagger}{\sqrt{2}}\right) \left(\frac{a - a^\dagger}{i\sqrt{2}} + P_0^*\right) , \\ K_0 &= \frac{a - a^\dagger}{i\sqrt{2}} + \frac{P_0 + P_0^*}{2} - \frac{1}{2} , \end{aligned} \tag{E17}$$

or,

$$\begin{aligned} K_- &= \left(\frac{a + a^\dagger}{\sqrt{2}} + P_0\right) \exp\left(\frac{a - a^\dagger}{\sqrt{2}}\right) , \\ K_+ &= \exp\left(-\frac{a - a^\dagger}{\sqrt{2}}\right) \left(\frac{a + a^\dagger}{\sqrt{2}} + P_0^*\right) , \\ K_0 &= \frac{a + a^\dagger}{\sqrt{2}} + \frac{P_0 + P_0^*}{2} - \frac{1}{2} . \end{aligned} \tag{E18}$$

The classical analog of this representation has been derived in [77].

E.2
Example

Let us consider now the following Hamiltonian:

$$\mathcal{H} = \varepsilon_k (b_k^\dagger b_k + b_{-k}^\dagger b_{-k}) + \Phi_2 b_k^\dagger b_{-k}^\dagger b_k b_{-k}$$
$$+ \Phi_1 (b_k^\dagger b_k^\dagger b_k b_k + b_{-k}^\dagger b_{-k}^\dagger b_{-k} b_{-k}) \, . \tag{E19}$$

Using the two mode representation of $SU(1,1)$ (E3), the Hamiltonian (E19) becomes:

$$\mathcal{H} = 2\Phi_1 - \varepsilon_k + (2\varepsilon_k - 6\Phi_1) K_0$$
$$+ 4\Phi_1 K_0^2 + (\Phi_2 - 2\Phi_1) K_+ K_- \, . \tag{E20}$$

Let us consider the pair states when the numbers of $\boldsymbol{k}$ and $-\boldsymbol{k}$ magnons are equal (formally, $b_k^\dagger b_k = b_{-k}^\dagger b_{-k}$). In this case the Casimir invariants (E12) and (E4) are equal to $-1/4$.

Using single-mode representation (E7), (E17), or (E18) in (E20), we can represent the two-mode Hamiltonian (E19) as the single-mode Hamiltonian with $a^\dagger$ and a, creation and annihilation Bose operators.

It is interesting that it is possible to reduce the problem of two nonlinear interacting magnons to the problem of a free-moving particle [201]. Substituting (E11) with

$$P_0 = \frac{3\Phi_1 + \Phi_2 - \varepsilon_k}{2\Phi_1 + \Phi_2} \tag{E21}$$

to (E20), we obtain the Hamiltonian

$$\mathcal{H} = \mathcal{H}_0 + \frac{P^2}{2m} \, . \tag{E22}$$

Here $\mathcal{H}_0 = -(\Phi_1 - \varepsilon_k)^2/(2\Phi_1 + \Phi_2)$ and $m = 1/[2(2\Phi_1 + \Phi_2)]$. If $2\Phi_1 + \Phi_2 > 0$, we have a ground state (condensate) with the energy $\mathcal{H}_0$ and an excitation with a quadratic spectrum of a particle of the mass m and momentum P.

Problem E.1. Consider mixed magnon pairs consisting of the following elements

$$\frac{1}{2}[c_k d_{-k} + c_{-k} d_k] \, ,$$
$$\frac{1}{2}\left[c_k^\dagger d_{-k}^\dagger + c_{-k}^\dagger d_k^\dagger\right] \, ,$$
$$\frac{1}{4}\left[c_k^\dagger c_k + c_{-k}^\dagger c_{-k}^\dagger + d_k^\dagger d_k + d_{-k}^\dagger d_{-k}^\dagger\right] + 1$$

from two different branches of spectra. Find the single-mode realization for this case.

Appendix F
Small Signal Amplification and Preventive Alarm Near the Onset of a Dynamic Instability

Near the onset of a dynamical instability, any time-periodic system can act to amplify small periodic perturbations [202, 203]. In this appendix we shall demonstrate that the modulation method described in Chapter 4 is a useful tool to study instabilities of the stationary state of parametrically excited magnons [204]. We shall show an amplification of small signals near the threshold of a so-called collective-acoustic instability in a system of parametric magnons as an example of this method. On the other hand, the effect of modulation response amplification can be considered as a preventive alarm which indicates that the strongly excited dynamic system is near an instability.

The collective-acoustic instability can occur at a given level of parametric excitation of magnetic system when the frequency Ω_0 of collective oscillations of magnon pairs approaches the frequency ω_e of elastic oscillations of the magnetic sample [205]. The mechanism of this instability is similar to a conventional oscillator with positive feedback: accidental elastic deformations of the sample via magnetoelastic interaction excite collective oscillations of the parametric magnon pairs. During the subsequent evolution of the collective oscillations the energy is transferred back from the magnetic to the elastic system and the transfer coefficient can be so large that self-excitation of sound takes place (the energy is "drawn" from the pumping field).

Let us consider a relatively simple mathematical model. The dynamic equations of excited parametric pairs contain just two variables: the number N_k of parametric magnons per magnetic cell and the phase θ_k of the magnon pair mismatching relative to the pump field:

$$\frac{d}{dt}\theta_k = \omega_p - 2\tilde{\omega}_k + 2hV_k \sin\theta_k - 2S_k N_k \,,$$
$$\frac{d}{dt}N_k = 2N_k\left(hV_k \sin\theta_k - \eta_k - \eta_{nl} N_k\right) \,. \qquad \text{(F1)}$$

At the steady state $d\theta_k/dt = dN_k/dt = 0$ we have $\omega_p - 2\tilde{\omega}_k = 0$ and θ_k and N_k as functions of overcriticality h/h_c. The damping parameter of magnons contains both linear (η_k) and nonlinear ($\eta_{nl} N_k$) terms, $h_c = \eta_k/V_k$ is the parallel pumping threshold, and V_k describes coupling with the pump field.

Nonequilibrium Magnons, First Edition. Vladimir L. Safonov.

The magnetoelastic interaction energy of interest can be represented in the form

$$\frac{\mathcal{H}_{\mathrm{me}}}{\hbar} = \sum_{q,q'} \Psi(q, q')(\mathfrak{b}_{e,q'} + \mathfrak{b}^{*}_{e,-q'}) b^{*}_{q} b_{q-q'} \,, \tag{F2}$$

where $\mathfrak{b}^{*}_{\mathrm{e},-q'}$, $\mathfrak{b}_{\mathrm{e}}$, q' and b^{*}_{q}, $b_{\mathrm{q\text{-}q'}}$ are the complex amplitudes of sound and spin waves, respectively. So far as the elastic vibration of the sample has negligibly small wave vector $q' \sim \pi/L$ (where L is the linear sample size) in comparison with the wave vectors of excited magnons, we can accept the following simple homogeneous approximation

$$\frac{\mathcal{H}_{me}}{\hbar} \simeq \Psi(\mathfrak{b}_e + \mathfrak{b}^{*}_e)\left(b^{*}_k b_k + b^{*}_{-k} b_{-k}\right) \,. \tag{F3}$$

Thus the dynamic equation for the elastic vibration can be represented as

$$\left(\frac{d}{dt} + \eta_{\mathrm{e}}\right) \mathfrak{b}_{\mathrm{e}} = -i\omega_{\mathrm{e}}\mathfrak{b}_{\mathrm{e}} - 2i\Psi N_k \,, \tag{F4}$$

where η_{e} is the elastic vibration relaxation rate. From this equation we can obviously see the mechanism of linear excitation of phonons by oscillations of N_k with the frequency close to ω_{e}. Deviations of N_k from the excited steady state are described in terms of collective oscillations (excitations similar to the second sound in superfluid helium).

On the other hand, the elastic vibrations $\mathfrak{b}_{\mathrm{e}} + \mathfrak{b}^{*}_{\mathrm{e}}$ modulate the frequency of the excited magnons as

$$\tilde{\omega}_k = \omega_k + 2\mathcal{T}_k N_k + \Psi(\mathfrak{b}_{\mathrm{e}} + \mathfrak{b}^{*}_{\mathrm{e}}) + \frac{\partial \omega_k}{\partial H} H_{\mathrm{m}} \cos \omega_{\mathrm{m}} t \tag{F5}$$

and excite collective oscillations. Here $\mathcal{T}_k$ and $\mathcal{S}_k$ (above) are the amplitudes of the four-magnon interaction. The formula (F5) is supplemented by the term $H_{\mathrm{m}} \cos \omega_{\mathrm{m}} t$, which corresponds to the external modulation field.

Let us consider the modulation response $\alpha_{\mathrm{m}} = \Delta P_{\mathrm{m}}/H_{\mathrm{m}}$, where ΔP_{m} is the modulation depth of the transmitted microwave power at the frequency ω_{m}. A system of magnons excited parametrically by an external microwave field of frequency ω_{p} absorbs part of the energy of this field. An applied RF magnetic field modulates the absorbed microwave power. In other words, a magnetic sample with parametrically excited magnons behaves as a nonlinear element or a modulator which mixes low- and high-frequency oscillations of the magnetic field. Solving the system of (F1), (F4) and (F5), we obtain the following expression for the modulation response:

$$\alpha_{\mathrm{m}} = 4\omega_{\mathrm{p}} \left|\frac{\partial \omega_k}{\partial H}\right| |\mathcal{S}_k| N_k^2 \left\{\left[(\eta_k + 2\eta_{\mathrm{nl}} N_k)\, A + \frac{\omega_{\mathrm{m}} B}{2}\right]^2 + \left[(\eta_k + 2\eta_{\mathrm{nl}} N_k)\, B - \frac{\omega_{\mathrm{m}} A}{2} - \frac{\omega_{\mathrm{m}}}{4\omega_{\mathrm{p}} \mathcal{S}_k N_k}\right]^2\right\}^{1/2} \,, \tag{F6}$$

where

$$A = \frac{A_1}{A_1^2 + B_1^2}\,, \qquad B = \frac{B_1}{A_1^2 + B_1^2}\,,$$
$$A_1 = \Omega_0^2 - \omega_{\rm m}^2 + 4(\eta_k + \eta_{\rm nl} N_k)\eta_{\rm nl} N_k - \rho_s\left(\eta_{\rm e}^2 + \omega_{\rm e}^2 - \omega_{\rm m}^2\right)\,,$$
$$B_1 = 2\omega_{\rm m}(\eta_k + 2\eta_{\rm nl} N_k + \eta_{\rm e}\rho_s)\,.$$

Here

$$\Omega_0 = 2N_k\left[\mathcal{S}_k\left(2\mathcal{T}_k + \mathcal{S}_k\right)\right]^{1/2} \qquad \text{(F7)}$$

is the frequency of collective oscillations (at $\eta_{\rm nl} = 0$) and

$$\rho_s = \frac{16\mathcal{S}_k(\Psi N_k)^2\omega_{\rm e}}{\left[\left(\eta_{\rm e}^2 + \omega_{\rm e}^2 - \omega_{\rm m}^2\right)^2 + 4\eta_{\rm e}^2\omega_{\rm m}^2\right]}\,.$$

We shall compare the modulation response (F6) with the experimental results obtained for parametrically excited nuclear magnons in a single crystal $CsMnF_3$ at temperature $T = 4.2$ K and magnetic field $H_0 = 0.77$ kOe (or, 61.3 kA/m). Pairs of nuclear magnons were excited by microwave parallel pumping of frequency $\omega_{\rm p}/2\pi = 806$ MHz.

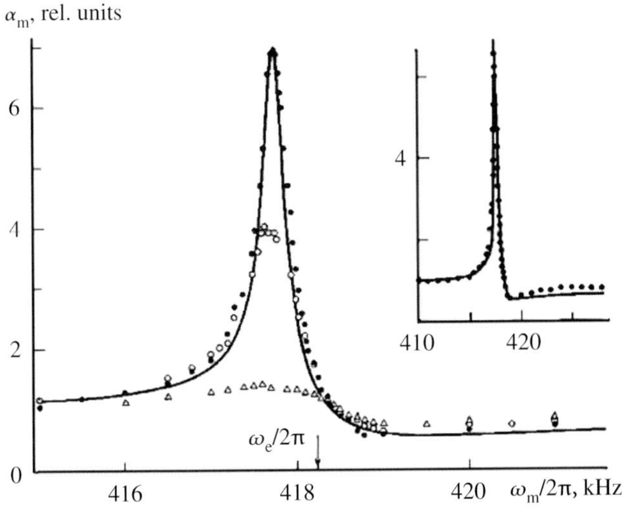

Figure F.1 Dependence of the modulation response $\alpha_{\rm m}$ on the modulation frequency $\omega_{\rm m}$ obtained for three different values of the pump amplitude h to the instability threshold $h_{\rm e}$: (black circles) $h/h_{\rm e} = 0.975 \pm 0.02$; (white circles) 0.95 ± 0.02; (white triangles) 0.79 ± 0.02. The continuous curve is calculated using (F6) with the following parameters: $\eta_{\rm e}/2\pi = 0.6$ kHz; $\eta_k/2\pi = 40$ kHz; $\eta_{\rm nl} \approx 0$; $\Psi/2\pi = 885$ kHz; $h/h_{\rm e} = 0.99$; $\Omega_0/2\pi = 389$ kHz. The inset shows the dependence of $\alpha_{\rm m}(\omega_{\rm m})$ obtained for $h/h_{\rm e} = 0.975$ in a wide range of frequencies $\omega_{\rm m}$. All the values of modulation response $\alpha_{\rm m}(\omega_{\rm m})$ are normalized to $\alpha_{\rm m}(\omega_{\rm m}/2\pi = 410\,\text{kHz})$ [204].

The collective-acoustic instability threshold h_e at the frequency $\omega_e/2\pi = 418.2\,\text{kHz}$ exceeded approximately threefold the parametric threshold amplitude: $h_e/h_c = 3$. The behavior of α_m was studied in the region of steady state of the parametric magnon system $h < h_e$. The dependencies of $\alpha_m(\omega_m)$ near ω_e for different subcriticalities h/h_e are plotted in Figure F.1.

The dependencies of $\alpha_m(h/h_e)$ were obtained at frequencies corresponding to the maximum and minimum of the modulation response near the elastic vibration frequency ω_e (see, Figure F.2).

The following details of the amplification effect should be mentioned.

a) The peak value of α_m strongly increases with the h approaching h_e. The maximum gain of the signal in this experiment was about $\sim 10^2$.
b) The peak was clearly asymmetric. It clearly indicates amplification of the modulation response at frequencies $\omega_m < \omega_e$ and the attenuation at higher frequencies.

The continuous curves in Figures F.1 and F.2 represent calculations of α_m using formula (F6) with the parameters that correspond to the experimental conditions. We see a good agreement between the experimental and theoretical results.

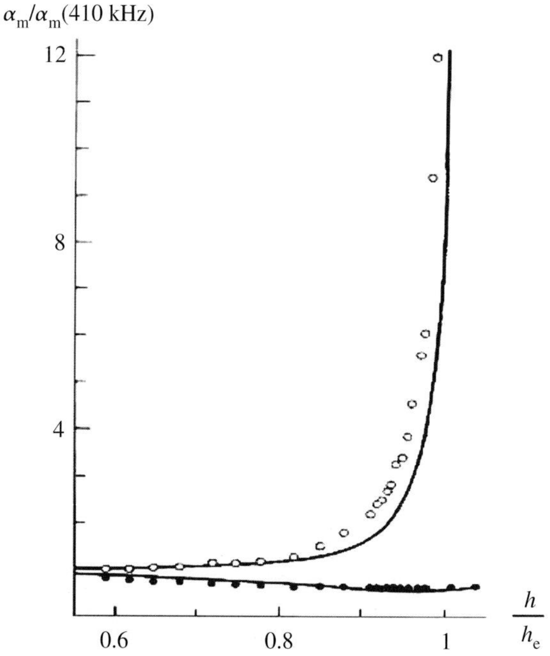

Figure F.2 Dependence of the relative modulation response α_m on the subcriticality h/h_e in the vicinity of the collective-acoustic instability for two frequencies: (white circles) maximum of $\alpha_m(\omega_m/2\pi = 417.8\,\text{kHz})$, (black circles) minimum of $\alpha_m(\omega_m/2\pi = 419.5\,\text{kHz})$. The continuous curves calculated with the use of (F6); all parameters are given in the text and in the caption of Figure F.1, in [204].

Note that the simple model describes not only the effect of amplification but also details of the effect.

A similar effect of small signal amplification near the threshold of a collective-acoustic instability was also observed and studied in a quite different system, the system of parametrically pumped quasi phonons [206]. By analogy, this method can be used in other excited nonlinear systems in the vicinity of instability.

Thus, the modulation method represents a powerful tool to study instabilities of the steady state in the excited system of magnon pairs. It can also be used as a preventive alarm near the onset of instability.

Problem F.1. Is it possible to find similar effects for mixed magnon pairs? Explain.

Appendix G
Noisy Pumping of Coherent Parametric Pairs

As we have already discussed in Chapter 4, the microwave resonator mode plays an important role in the process of parametric excitation of magnon pairs. Suppose we apply to the sample in the resonator cavity two independent microwave pumping fields:

$$F(t) = F\left[\exp(-i\omega_{\mathrm{p1}}t) + \exp(-i\omega_{\mathrm{p2}}t)\right] . \tag{G1}$$

Let the frequency of the first pump field be equal to the resonator frequency $\omega_{\mathrm{p1}} = \omega_{\mathrm{R}}$, and the frequency of the second pump field be slightly different $\omega_{\mathrm{p2}} = \omega_{\mathrm{R}} + \Delta\omega$. In this case the parametric threshold of the first pump field is lower than the threshold of the second one. The excited parametric pairs would absorb both the resonance microwave fields and the energy of the second field, which is slightly out of resonance with the pairs. In addition, the nonlinear radiation damping that appears due to excited state would suppress the process of parametric pairs excitation related to the second pump field. One can expect such a "competition" between parametric pairs for more complex microwave pumping field.

In this appendix we examine the behavior of parametrically excited pairs in the case of a noisy pump. Our purpose is to demonstrate that incoherent electromagnetic oscillations in a resonator with the sample can be effectively transformed into a coherent magnetic oscillation of a parametric pair. Efficiency of 50 % and a frequency narrowing (squeezing) of more than a factor of 10 have been experimentally demonstrated using the system of quasi phonons in $FeBO_3$. The effect was observed for quasi phonons [207] and nuclear magnons [208, 209]. It is expected to be general and to occur in a wide class of magnetic materials over a wide range of microwave frequencies. This effect provides the possibility of studying and developing a new class of microwave magnetic devices that transform incoherent microwaves and noise into coherent microwave signals.

Nonequilibrium Magnons, First Edition. Vladimir L. Safonov.

G.1
Experimental Procedure

A noisy field in the frequency interval $\omega_p - \Delta\omega/2$, $\omega_p + \Delta\omega/2$ with $\omega_p/2 =$ 1200–1250 MHz and $\Delta\omega = 0.3$ MHz was produced in antiferromagnetic $FeBO_3$ to excite quasi phonon pairs.

The noise RF signal from a generator was used for frequency modulation of a microwave source. The pump spectrum was studied with the spectrum analyzer and was found to be approximately Gaussian. The spectral width $\Delta\omega$ was measured within 10 % at half-maximum.

A sample with a volume of $1.2 \times 3.5 \times 7\,\text{mm}^3$ was placed in the helix microwave resonator, which had a low quality factor ~ 300, so that the edge of the spectrum of the noisy pump would not be cut off. Experiments were carried out at $T = 77$ and 293 K in a field $H = 80$–400 Oe. The static magnetic field and all the alternating magnetic fields were parallel. The relaxation rate of the quasi phonons which were excited was $\eta_k \simeq 2\pi \cdot 0.1$ MHz.

The microwave pumping was carried out in a pulsed regime with a repetition frequency of 10–500 Hz and a pulse length of 20–1000 μs. On the pump pulses which were transmitted through the cavity and detected we observed the onset of a nonlinear absorption above a certain threshold power. The relative error of the measurements of this threshold was less than 5.

To observe the nonequilibrium Bose condensate we used two methods, based on the observation of collective effects arising from this state.

The first method was to observe the modulation response α_m. This method was developed for, and has been widely used for, studying the properties of a nonequilibrium Bose condensate of magnons, which arises in the case of monochromatic pumping. A weak RF magnetic field $H_m \cos \omega_m t$ was applied to the sample; this weak field modulated the spectrum of quasi phonons. In this case, oscillations of the amplitude and phase of the nonequilibrium Bose condensate around their equilibrium values arise (these are "collective oscillations"). They lead to a modulation of the microwave power absorbed by the sample, with an amplitude ΔP and a frequency ω_m. The onset of this amplitude modulation ($\Delta P = \alpha_m H_m$) indicates the existence of a nonequilibrium Bose condensate in the sample. To find α_m, we detected the microwave signal transmitted through the cavity and sent it to a tuned microvoltmeter, tuned to the frequency ω_m. This signal was then sent to a lock-in detector; the output signal from the detector was fed to the Y-input of an x, y-chart recorder.

The second method was to observe electromagnetic radiation from the system of excited quasi phonons. The effect, which we observed just recently (and which will be described in detail in a separate paper), is as follows: after the pump is turned off, the coherent parametric pair produced by the monochromatic microwave pump causes characteristic electromagnetic radiation from the sample. The intensity of this radiation is not monotonic in time; its frequency is close to the pump frequency.

G.2
Results and Discussion

This study showed that the picture of the nonlinear absorption of microwave power is qualitatively the same in the cases of noisy pumps produced by frequency modulation and by phase modulation (a noisy modulation of the spectrum). One should distinguish two thresholds, $h_c^{(1)} < h_c^{(2)}$: the first corresponds to the onset of nonlinear microwave absorption, and the second to the formation of a nonequilibrium Bose condensate.

Figure G.1 shows a chart recording of the amplitude of the modulation response signal versus the total pump power. Also shown are oscilloscope traces of the microwave signal transmitted through the cavity. One can see that there is no nonlinear absorption at point A. On trace B there are some "spikes" of nonlinear absorption above the threshold $h_c^{(1)}$. With increasing pump power, this absorption increases on average (trace C), but there is still no modulation response signal, although nonequilibrium quasi phonons do exist (on average) in the sample. As the microwave power is raised further, at $h > h_c^{(2)}$, we observe a characteristic decay, as in the case of a monochromatic pump. At the same time, a modulation response arises, signifying the presence (on average) of phase correlations in the system, that is, signifying the formation of a nonequilibrium Bose condensate. The ratio of thresholds, $h_c^{(2)}/h_c^{(1)}$, increases with increasing width $\Delta\omega$ and reaches a value of 2. The ratio of $h_c^{(2)}$ to the threshold in the absence of noise reaches $h_c^{(2)}/h_{c0} = 5$.

The formation of a nonequilibrium Bose condensate above the threshold $h_c^{(2)}$ was also detected from the electromagnetic emission from the sample, observed after the end of the pump pulse (see, Figure G.2). We wish to emphasize that at $h < h_c^{(2)}$ there is no such radiation, while above the threshold $h_c^{(2)}$ the radiation has approximately the same intensity and the same induction-signal decay time

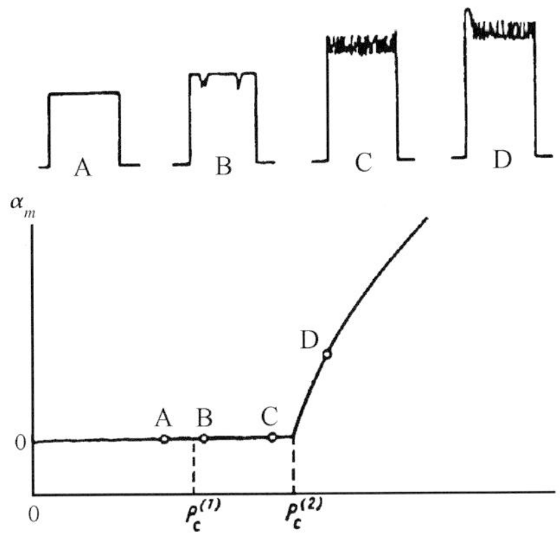

Figure G.1 Modulation response signal α_m versus the pump power $P \propto h^2$ and oscilloscope traces of the microwave pulses transmitted through the resonator at points (A), (B), (C), and (D). The pulse length is 100 μs and the amplitude of the modulation field $H_m < 0.1$ Oe (in SI: 7.96 A/m) [207].

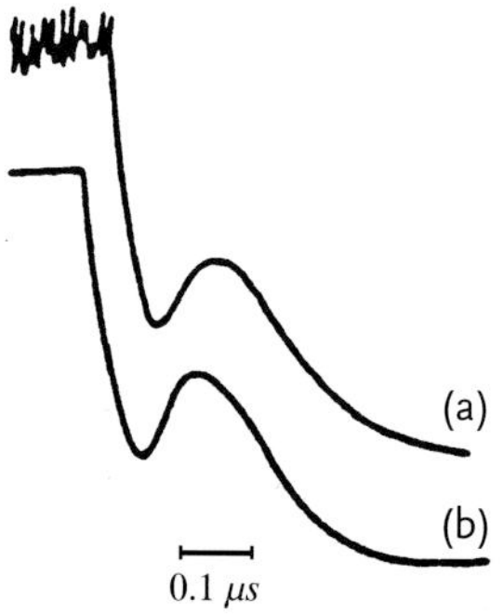

Figure G.2 Oscilloscope traces of the trailing edge of the microwave pulses, along with a signal representing the electromagnetic radiation from the sample: noisy pump (a), and monochromatic pump (b) [207].

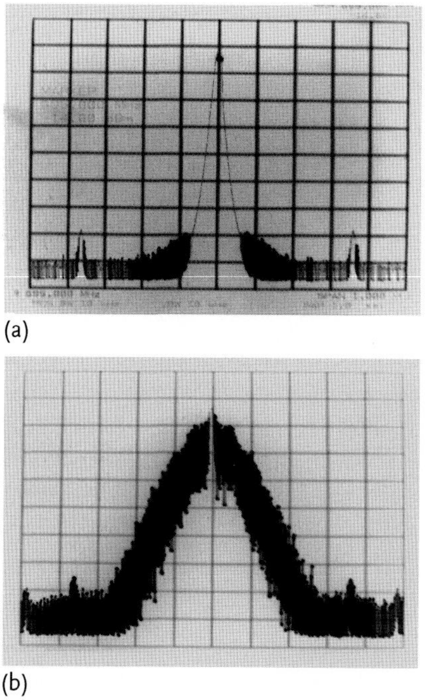

Figure G.3 Typical spectrum analyzer images showing (a) coherent pumping, and (b) incoherent pumping. Small peaks correspond to the modulation response. Pumping frequencies range is 0.8–1.5 GHz, modulation frequencies are 0.1–1 MHz. All measurements were performed at room temperature.

as in the case of the emission of a parametric pair excited by a monochromatic microwave field. It can thus be concluded that the condensate of quasi phonons which arises in the case of a noisy pump has approximately the same frequency width as in the case of a monochromatic pump.

Thus, experimentally it was demonstrated that incoherent microwaves are absorbed by the magnetic system and transformed into a coherent microwave signal with high efficiency (Figure G.3).

Problem G.1. Simulate the process of parametric excitation of N parametric pairs by a noisy field in a resonator. For simplicity assume that all pairs have no interaction with each other and have the same damping parameter.

G.2.1 Discussion

a) Explain the role of interactions between different parametric pairs.
b) Is coexistence of noisy pumping in a resonator and Bose–Einstein condensation of quasi equilibrium magnons possible?
c) Can the uniform magnetization precession play the role of resonator mode? If yes, explain.

References

1 Feynman, R.P., Leighton, R.B., and Sands, M. (1966) *The Feynman Lectures on Physics*, Addison-Wesley, Reading.

2 Crawford, F.S. (1968) *Waves*, Berkeley Physics Course, McGraw-Hill, New York.

3 Rabinovich, M.I. and Trubetskov, D.I. (1989) *Oscillations and Waves in Linear and Nonlinear Systems*, Kluwer Academic Publishers, Dordrecht, Boston, London.

4 Zakharov, V.E., L'vov, V.S., and Falkovich, G. (1992) *Kolmogorov Spectra of Turbulence. 1. Wave Turbulence*, Springer-Verlag, Berlin.

5 Scully, M.O. and Zubairy, M.S. (1997) *Quantum Optics*, Cambridge University Press, Cambridge.

6 Landau, L.D. and Lifshitz, E.M. (1976) *Mechanics*, 3rd edn, Course of Theoretical Physics, Reed Educational and Professional Publishing Ltd, Oxford.

7 Bloch, F. (1930) Zur Theorie des Ferromagnetismus. *Zeitschrift für Physik*, **49**, 206–219.

8 Holstein, T. and Primakoff, H. (1940) Field dependence of the intrinsic domain magnetization of a ferromagnet. *Physical Review*, **58**, 1098–1113.

9 Akhiezer, A.I., Bar'yakhtar, V.G., and Peletminskii, S.V. (1968) *Spin Waves*, North-Holland, Amsterdam; Interscience (Wiley), New York.

10 Gurevich, A.G. and Melkov, G.A. (1996) *Magnetization Oscillations and Waves*, CRC Press, Inc., Boca Raton.

11 Mattis, D.C. (1981) *The Theory of Magnetism I*, Springer, New York.

12 White, R.M. (2007) *Quantum Theory of Magnetism*, 3rd edn, Springer, Berlin.

13 Stancil, D.D. and Prabhakar A. (2009) *Spin Waves. Theory and Applications*, Springer, Berlin.

14 Landau, L. and Lifshitz, E. (1935) On the theory of the dispersion of magnetic permeability in ferromagnetic bodies. *Phys. Z. Sowjetunion*, **8**, 153; (1967) in Landau L.D. *Collected Papers*, ed D. ter Haar, Gordon and Breach, New York, 101.

15 Kotyuzhanskii, B.Ya. and Prozorova, L.A. (1983) Change of magnetization following parametric excitation of magnons in antiferromagnetic $FeBO_3$. *Zhurnal Eksperimentalnoi i Teoreticheskoi Fiziki*, **85** (4), 1461–1464; [*Soviet Physics JETP*, **58** (4), 846–848].

16 Kaganov, M.I., Pustylnik, N.B., and Shalaeva, T.I. (1997) Magnons, magnetic polaritons, magnetostatic waves. *Uspekhi Fizicheskikh Nauk*, **167** (2), 191–237; [*Physics Uspekhi*, **40** (2), 181–224].

17 Chikazumi, S. (1997) *Physics of Ferromagnetism*, 2nd edn, Oxford University Press, Oxford.

18 du Trémolet de Lacheisserie, E., Gignoux, D., and Schlenker M. (eds) (2005) *Magnetism Fundamentals*, Springer Science + Business Media Inc., Boston.

19 Nolting, W. and Ramakanth, A. (2007) *Quantum Theory of Magnetism*, Springer-Verlag, Berlin.

20 Turov, E.A. (1965) *Physical Properties of Magnetically Ordered Crystals*, Academic Press, New York.

21 Andreeev, A.F. and Marchenko, V.I. (1980) Symmetry and the macroscopic dynamics of of magnetic materials. *Us-*

Nonequilibrium Magnons, First Edition. Vladimir L. Safonov.

pekhi Fizicheskikh Nauk, **130**, 39–63. [*Soviet Physics Uspekhi*, **23** (1), 21–34].
22 Landau, L.D. and Lifshitz, E.M. (1984) *Electrodynamics of Continuous Media*, 2nd edn, Course of Theoretical Physics, Pergamon Press, Oxford.
23 Lutovinov, V.S. and Safonov, V.L. (1980) Effect of dipole–dipole interaction on the parametric excitation of magnons in easy-plane antiferromagnets. *Fizika Tverdogo Tela (Leningrad)*, **22** (9), 2640–2650 [*Soviet Physics Solid State*, **22** (9), 1541–1546].
24 Breus, S.A. and Sobolev, V.L. (1980) On magnon damping in antiferromagnets in zero external magnetic fields, *Physica Status Solidi (b)*, **98** (2), 757–764.
25 Svitsov, L.E. (1991) Change in magnetic moment of the easy-plane antiferromagnet $MnCO_3$ by parametric excitation of magnons. *Zhurnal Eksperimentalnoi i Teoreticheskoi Fiziki*, **99** (5), 1612–1618; [*Soviet Physics JETP*, **72** (5), 900–903].
26 Krivoruchko, V.N. and Yablonskii, D.A. (1981) On the problem of calculating the Bogolyubov U-V transformation coefficients. *Physica Status Solidi (b)*, **103**, K41–K45.
27 Safonov, V.L. (1984) Interactions of magnons in the six-sublattice antiferromagnet. *Physica Status Solidi (b)*, **126**, 197–206.
28 Kolokolov, I.V., L'vov, V.S., and Cherepanov, V.B. (1983) Spin wave spectra and thermodynamics of yttrium-iron garnet – a twenty-sublattice ferrimagnet. *Zhurnal Eksperimentalnoi i Teoreticheskoi Fiziki*, **84** (3), 1041–1063; [*Soviet Physics JETP*, **57** (3), 605–613].
29 Cherepanov, V., Kolokolov, I., and L'vov, V. (1993) The saga of YIG: spectra, thermodynamics, interaction and relaxation of magnons in a complex magnet, *Physics Reports*, **229** (3), 81–144.
30 de Gennes, P.G., Pincus, P.A., Hartmann-Boutron, F., and Winter, J.M. (1963) Nuclear magnetic resonance modes in magnetic material. *Physical Review*, **129** (3), 1105–1115.
31 Adams B.T., Hinderks L.W., and Richards P.M. (1970) Direct excitation of two nuclear spin waves by parallel pumping in $CsMnF_3$, *Journal of Applied Physics*, **41** (3), 931–932.
32 Turov, E.A. and Petrov, M.P. (1972) *Nuclear Magnetic Resonance in Ferro- and Antiferromagnets*, Israel Program for Scientific Translations, Jerusalem.
33 Tulin, V.A. (1979) Nuclear spin waves in magnetically ordered substances. *Fizika Nizkikh Temperatur, 5*, 965; [*Soviet Journal Low Temperature Physics*, **5**, 455].
34 Andrienko, A.V., Ozhogin, V.I., Safonov, V.L., and Yakubovskii, A.Yu. (1991) Nuclear spin wave research. *Uspekhi Fizicheskikh Nauk*, **161** (10), 1–35; [*Soviet Physics Uspekhi*, **34** (10), 843–861].
35 Suhl H. (1958) Effective nuclear spin interactions in ferromagnets. *Physical Review*, **109**, 606.
36 Nakamura T. (1958) Indirect coupling of nuclear spins in antiferromagnet with particular reference to MnF_2 at very low temperature. *Progress of Theoretical Physics*, **20**, 542–552.
37 Safonov, V.L. (1988) Phase transition in system of nuclear spins with indirect Suhl–Nakamura interaction. *Zhurnal Eksperimentalnoi i Teoreticheskoi Fiziki*, **94** (11), 263–270; [*Soviet Physics JETP, 67* (11), 2324–2327].
38 Svistov, L.E., Löw, J., and Benner, H. (1993) The effect of nuclear spin waves on the magnetization of $MnCO_3$. *Journal of Physics: Condensed Matter*, **5**, 4215–4224.
39 Ozhogin, V.I. and Preobrazhenskii, V.L. (1988) Uspekhi Fizicheskikh Nauk, **155** (4), 593–621; [*Soviet Physics Uspekhi*, **31**, 713–729].
40 Svistov, L.E., Safonov, V.L., and Khachevatskaya, C.R. (1997) Spin-wave resonances in nonuniformly strained films of $FeBO_3$. *Zhurnal Eksperimentalnoi i Teoreticheskoi Fiziki*, **112** (2), 564–573; [*Journal of Experimental and Theoretical Physics*, **85** (2), 307–312].
41 Landau, L.D. and Lifshitz, E.M. (1970) *Theory of Elasticity*, 2nd edn, Course of Theoretical Physics, Pergamon Press, Oxford.
42 Lebedev, A.Yu., Ozhogin, V.I., Safonov, V.L., and Yakubovskii, A.Yu. (1983) Nonlinear magnetoacoustics of orthoferrite near spin flip. *Zhurnal Eksperimentalnoi*

i Teoreticheskoi Fiziki, **85** (3), 1059–1071; [*Sov. Phys. JETP*, **58** (3), 616–623].

43 Svistov, L.E., Safonov, V.L., Löw, J., and Benner, H. (1994) Detection of UHF sound in the antiferromagnet $FeBO_3$ by a SQUID magnetometer. *Journal of Physics: Condensed Matter* **6**, 8051–8063.

44 Zhang, Q., Mino, M., Safonov, V.L., and Yamazaki, H. (2000) (2000). Microwave radiation of parametrically excited quasiphonons in antiferromagnet $FeBO_3$. *Journal of Physical Society of Japan*, **69**, 41–44.

45 Lax, M. (1966) Quantum noise. IV. Quantum theory of noise sources. *Physical Review*, **145** (1), 110–145.

46 Lax, M. (1968) Fluctuation and coherence phenomena in classical and quantum physics, in *Statistical Physics, Phase Transitions, and Superconductivity* (eds M. Chrétien, E.P. Gross, and S. Deśer), Gordon and Breach, Science Publishers, Inc., New York, 269–468.

47 Sparks, M. (1964) *Ferromagnetic-Relaxation Theory* , McGraw-Hill, New York.

48 Woolsey, R.B. and White R.M. (1969) Theory of spin-wave relaxation in antiferromagnets. *Physical Review*, **188** (2), 813–820.

49 Andrienko A.V. and Poddyakov L.V. (1987) Relaxation of electron spin waves in antiferromagnetic $CsMnF_3$. *Zhurnal Eksperimentalnoi i Teoreticheskoi Fiziki*, **93** (5),1848–1853; [*Soviet Physics JETP*, **66** (5), 1055–1057].

50 Kambersky, V. (1970) On the Landau–Lifshitz relaxation in ferromagnetic metals. *Canadian Journal of Physics*, **48**, 2906–2911.

51 Kambersky, V. and Patton, C.E. (1975) Spin-wave relaxation and phenomenological damping in ferromagnetic resonance. *Physical Review B*, **11**, 2668–2672.

52 Mikhailov, A.S. and Farzetdinova, R.M. (1981) Quantum theory of spin-wave relaxation by two-level impurities. *Zhurnal Eksperimentalnoi i Teoreticheskoi Fiziki*, **80** (4), 1524–1538. [*Soviet Physics JETP*, **53** (4), 782–789].

53 Mikhailov, A.S. and Farzetdinova, R.M. (1983) Processes of spin-wave relaxation on paramagnetic impurities in antiferromagnets. *Zhurnal Eksperimentalnoi i Teoreticheskoi Fiziki*, **84** (1), 190–204. [*Soviet Physics JETP*, **57** (1), 109–116].

54 Safonov, V.L. and Farzetdinova, R.M. (1991) Slow relaxation of Bose-quasiparticles on multilevel systems. *Physica Status Solidi (b)*, **163**, 259–266.

55 Safonov, V.L. and H.N. Bertram (2003) Linear stochastic magnetization dynamics and microscopic relaxation mechanisms. *Journal of Applied Physics*, **94**, 529–538.

56 Patton, C.E. (1968) *Journal of Applied Physics*, **39**, 3060.

57 Patton, C.E. (1975) Microwave Resonance and Relaxation, in *Magnetic Oxides*, (ed by D.J. Craik), John Wiley & Sons Ltd, London, 575.

58 Patton, C.E. and Wilts, C.H. (1967) *Journal of Applied Physics*, **38**, 3537.

59 Patton, C.E., Frait, Z., and Wilts, C.H. (1975) *Journal of Applied Physics*, **46**, 5002.

60 Krinchik, G.S. (1985) *Physics of Magnetic Phenomena* (in Russian), Moscow State University, Moscow.

61 Suhl, H. (1998) Theory of the magnetic damping constant. *IEEE Transactions on Magnetics*, **34**, 1834–1838.

62 Gilbert, T.L. (1955) *Physical Review*, **100**, 1243.

63 Gilbert, T.L. (2004) A phenomenological theory of damping in ferromagnetic materials, *IEEE Transactions on Magnetics*, **40** (6), 3443–3449.

64 Safonov, V.L. (2002) Tensor form of magnetization damping. *Journal of Applied Physics*, **91**, 8653–8655.

65 Andrienko, A.V., Ozhogin, V.I., Safonov, V.L., and Yakubovskii, A.Yu. (1985) Influence of the electronic-magnon relaxation rate on the damping of nuclear spin waves in antiferromagnets. *Zhurnal Eksperimentalnoi i Teoreticheskoi Fiziki* **89** (4), 1371–1381; [*Soviet Physics JETP* **62** (4), 794–799].

66 Seavey, M.H. (1969) Magnon-phonon interaction and determination of exchange constants in $CsMnF_3$ *Physical Review Letters*, **23** (3), 132–135.

67 Safonov, V.L., Loaiza, P.M., and Svistov, L.E. (1997) Relaxation of magnetoelastic vibrations in antiferromagnetic $FeBO_3$.

Journal of Magnetism and Magnetic Materials, **173**, 43–50.

68 Kotyuzhanskii, B.Ya. and Prozorova, L.A. (1981) Parametric excitation of spin waves in the antiferromagnet $FeBO_3$. *Zhurnal Eksperimentalnoi i Teoreticheskoi Fiziki* **81** (5), 1913–1924; [*Soviet Physics JETP* **54** (5), 1013–1019].

69 Andrienko, A.V. (2000) Investigation of the Parametric Excitation and Relaxation of Phonons in Antiferromagnetic α-Fe_2O_3 *Zhurnal Eksperimentalnoi i Teoreticheskoi Fiziki*, **117** (3), 582–592; [*Journal of Experimental and Theoretical Physics*, **90** (3), 508–516].

70 Suhl, H. (1957) The theory of ferromagnetic resonance at high signal powers. *Journal of Physics and Chemistry of Solids*, **1**, 209–227.

71 Schlömann, E. (1959) Fine structure in the decline of the ferromagnetic resonance absorption with increasing power level. *Physical Review*, **116** (4), 828–837.

72 Zakharov, V.E., L'vov, V.S., and Starobinets, S.S. (1974) Spin-wave turbulence beyond the parametric excitation threshold. *Uspeckhi Fizicheskikh Nauk*, **114**, 609–654; [(1975) *Soviet Physics Uspekhi* **18**, 896–919].

73 L'vov, V.S. (1994) *Wave Turbulence under Parametric Excitation*, Springer-Verlag, Berlin.

74 Safonov, V.L. (1991) Influence of the microwave cavity mode relaxation on the process of parametric magnons excitation. *Journal of Magnetism and Magnetic Materials*, **97**, L1–L3.

75 Safonov, V.L. and Yamazaki, H. (1996) Nonlinear radiation damping of parametrically excited spin waves. *Journal of Magnetism and Magnetic Materials*, **161**, 275–281.

76 Ozhogin, V.I. and Yakubovskii, A.Yu. (1974) Parametric pairs in antiferromagnet with easy plane anisotropy. *Zhurnal Eksperimentalnoi i Teoreticheskoi Fiziki* **67** (1), 287–308; [*Soviet Physics JETP* **40** (1), 145–153].

77 Safonov, V.L., Shi, Q., Mino, M., and Yamazaki, H. (1997) Parametric pair of waves as a nonlinear oscillator. *Journal of the Physical Society of Japan*, **66** (7), 1916–1919.

78 Bryant P.H., Jeffries C.D., and Nakamura K. (1988) Spin-wave dynamics in a ferromagnetic sphere. *Physical Review A*, **38** (8), 4223–4240.

79 Andrienko, A.V. and Podd'yakov, L.V. (1989) Parallel pumping of phonons in antiferromagnetic $FeBO_3$ by a microwave magnetic field. *Zhurnal Eksperimentalnoi i Teoreticheskoi Fiziki*, **95** (6), 2117–2124; [*Soviet Physics JETP*, **68** (6), 1224–1228].

80 Andrienko, A.V., Podd'yakov, L.V., and Safonov V.L. (1992) Parametric excitation of magnetoelastic waves in single crystals of $CoCO_3$ and $FeBO_3$. *Zhurnal Eksperimentalnoi i Teoreticheskoi Fiziki*, **101** (3), 1083–1098; [*Soviet Physics JETP*, **74** (3), 579–587].

81 Andrienko, A.V. and Safonov, V.L. (1994) Electromagnetic emission by a system of nonequilibrium quasiphonons in an antiferromagnet. *Pisma Zhurnal Eksperimentalnoi i Teoreticheskoi Fiziki*, **60** (11), 787–791; [*JETP Letters*, **60** (11), 800–804].

82 Andrienko, A.V. and Safonov, V.L. (1995) Observation of coupled photon-phonon oscillations with parametric excitation of magnetoelastic waves in an antiferromagnet. *Pisma Zhurnal Eksperimentalnoi i Teoreticheskoi Fiziki*, **62** (2), 147–151; [*JETP Letters*, **62** (2), 162–167].

83 Andrienko, A.V., Safonov, V.L., and Yamazaki, H. (1998) Study of parallel pumping of magnetoelastic waves in an antiferromagnetic $FeBO_3$. *Journal of Physical Society Japan*, **67**, 2893–2903.

84 Poole, Ch.P. (1967) *Electron Spin Resonance. A Comprehensive Treatise on Experimental Techniques*. John Wiley & Sons Inc., New York.

85 Andrienko, A.V., Ozhogin, V.I., Safonov, V.L., and Yakubovskii, A.Yu. (1983) Modulation method of investigating spin waves beyond the parametric-excitation threshold. *Zhurnal Eksperimentalnoi i Teoreticheskoi Fiziki*, **84** (4), 1474–1480; [*Soviet Physics JETP*, **57** (4), 858–862].

86 Ozhogin, V.I., Andrienko, A.V., Safonov, V.L., and Yakubovskii, A.Yu. (1986) Measurement of modulation response as a new method for study parametric spin waves. *Journal of Magnetism and Magnetic Materials*, **54–57**, 1151–1153.

87 Andrienko, A.V., Safonov, V.L., and Yakubovskii, A.Yu. (1987) Modulation-method investigation of the steady state of parametric spin waves in antiferromagnets. *Zhurnal Eksperimentalnoi i Teoreticheskoi Fiziki*, **93** (3), 907–917; [*Soviet Physics JETP*, **66** (3), 511–516].

88 Kveder, V.V. and Prozorova, L.A. (1974) Investigation of the beyond-threshold susceptibility in antiferromagnetic $MnCO_3$ and $CsMnF_3$ in parametric excitation of spin waves. *Pisma v Zhurnal Eksperimentalnoi i Teoreticheskoi Fiziki*, **19** (11), 683–686; [*JETP Letters*, **19** (11), 353–354].

89 Shi, Q., Mino, M., Safonov, V.L., and Yamazaki H. (1998) Observation of nonlinear coupled photon-magnon oscillations in YIG. *Journal of Physical Society Japan*, **67** (9), 3283–3287.

90 G. Wiese, G., Krug von Nidda, H.-A., and Benner, H. (1991) Temperature-induced nonlinearity at parametrically excited spin waves. *Europhysics Letters*, **15**, 585–588.

91 Sharaevskii, Yu.P., Grishin, V.S., Gurzo, V.V., Derunov, A.V., and Shakhat, A.A. (1995) The interaction of regular and noisy signals in nonlinear MSW line [in Russian] *Radiotekhnika i Elektronika*, **40** (7), 1064–1068.

92 Hinderks, L.W. and Richards, P.M. (1968) Excitation of nuclear and electronic spin waves in $RbMnF_3$ by parallel pumping. *Journal of Applied Physics*, **39** (2), 824–825.

93 Andrienko, A.V., Safonov, V.L., and Yakubovskii, A.Yu. (1989) Investigation of a combined resonance of mixed parametric magnon pairs in antiferromagnet $MnCO_3$. *Zhurnal Eksperimentalnoi i Teoreticheskoi Fiziki*, **96** (2), 641–654; [*Soviet Physics JETP*, **69** (2), 363–370].

94 Morozov, V.G. and Mukhai, A.N. (1982) Kinetic description of collective oscillations in a system of parametric spin waves. I. Kinetic equations. Homogeneous collective modes. *Teoreticheskaya i Matematicheskaya Fizika*, **51** (2), 234–246; [*Theoretical and Mathematical Physics*, **51** (2), 468–476].

95 Morozov, V.G. and Mukhai, A.N. (1982) Kinetic description of collective oscillations in a system of parametric spin waves II. Inhomogeneous collective modes. *Teoreticheskaya i Matematicheskaya Fizika*, **53** (2), 305–317; [*Theoretical and Mathematical Physics*, **53** (2), 1142–1149].

96 Morozov, V.G. and Mukhai, A.N. (1992) Fokker–Planck method in the theory of parametric resonance of spin waves. *Teoreticheskaya i Matematicheskaya Fizika*, **90** (2), 278–300; [*Theoretical and Mathematical Physics*, **90** (2), 189–204].

97 Goldman, M. (1970) *Spin Temperature and Nuclear Magnetic Resonance in Solids*, Clarendon Press, Oxford.

98 Slichter, C.P. (1980) *Principles of Magnetic Resonance*, Springer-Verlag, Berlin.

99 Bogolyubov, N.N. (1959) The compensation principle and the self-consistent field method. *Uspekhi Fizicheskikh Nauk*, **67** (4), 549–580; [*Soviet Physics Uspekhi* **2** (4), 236–254].

100 Landau, L.D. and Lifshitz, E.M. (1980) *Statistical Physics, Part 1*, Butterworth-Heinemann, Oxford.

101 Kalafati, Yu.D. and Safonov, V.L. (1989) Thermodynamic theory of saturation of the impulse magnon excitation. *Journal de Physique*, **50**, 1157–1161.

102 Kalafati, Yu.D. and Safonov, V.L. (1989) Thermodynamic approach in the theory of parametric resonance of magnons. *Zhurnal Experimentalnoi i Teoreticheskoi Fiziki*, **95** (6), 2009–2020; [*Soviet Physics JETP*, **68** (6), 1162–1167].

103 Kalafati, Yu.D. and Safonov, V.L. (1991) Theory of quasiequilibrium effects in a systems of magnons excited by incoherent pumping. *Zhurnal Eksperimentalnoi i Teoreticheskoi Fiziki*, **100** (5), 1511–1521; [*Soviet Physics JETP*, **73** (5), 836–841].

104 Lavrinenko, A.V., L'vov, V.S., Melkov, G.A., and Cherepanov, V.B. (1981) "Kinetic instability" of a strongly nonequilibrium system of spin waves and tunable radiation of a ferrite. *Zhurnal Eksperimentalnoi i Teoreticheskoi Fiziki*, **81**, 1022–1036; [*Soviet Physics JETP*, **54** (3), 542–549].

105 Melkov, G.A., Safonov, V.L., Taranenko, A.Yu., and Sholom, S.V. (1994) Kinetic instability and Bose condensation of nonequilibrium magnons. *Journal of*

Magnetism and Magnetic Materials, **132**, 184.

106 Govorkov, S.A. and Tulin, V.A. (1989) Heating and energy diffusion in a coupled electron-nuclear magnetic system of antiferromagnetic $CsMnF_3$ subjected to strong parametric excitation. *Zhurnal Eksperimentalnoi i Teoreticheskoi Fiziki*, **95** (4), 1398–1403; [*Soviet Physics JETP*, **68** (4), 807–810].

107 Demokritov, S.O., Demidov, V.E., Dzyapko, O., Melkov, G.A., Serga, A.A., Hillebrands, B., and Slavin, A.N. (2006) Bose–Einstein condensation of quasi-equilibrium magnons at room temperature under pumping. *Nature*, **443**, 430–433.

108 Dzyapko, O., Demidov, V.E., Demokritov, S.O., Melkov, G.A., and Safonov, V.L. (2008) Monochromatic microwave radiation from the system of strongly excited magnons. *Applied Physics Letters*, **92**, #162 510 3p.

109 Demidov, V.E., Dzyapko, O., Demokritov, S.O., Melkov, G.A., and Slavin, A.N. (2007) Thermalization of a parametrically driven magnon gas leading to Bose–Einstein condensation. *Physical Review Letters*, **99**, # 037 205 4p.

110 Safonov, V.L. (1992) Thermodynamic description of strongly excited Bose systems. *Physica A*, **188**, 675–686.

111 Pitaevskii, L.P. and Lifshitz, E.M. (1980) *Statistical Physics, Part 2*, Butterworth-Heinemann, Oxford.

112 Pitaevskii, L. and Stringari, S. (2003) *Bose–Einstein Condensation*, Clarendon Press, Oxford.

113 Anglin, J.R. and Ketterle, W. (2002) Bose–Einstein condensation of atomic gases. *Nature*, **416**, 211–218.

114 Pethick, C.J. and Smith, H. (2002) *Bose–Einstein Condensation in Dilute Gases*, Cambridge University Press, Cambridge.

115 Martellucci, S., Chester, A.N., Aspect, A., and Inguscio, M. (eds) (2002) *Bose–Einstein Condensates and Atom Lasers*, Kluwer Academic Publishers, New York.

116 Gunton, J.D. and Buckingham, M.J. (1968) Condensation of the ideal Bose gas as a cooperative transition. *Physical Review*, **166** (1), 152–158.

117 Batyev, A.G. and Braginskii, L.S. (1984) Antiferromagnet in a strong magnetic field: Analogy with Bose gas. *Zhurnal Eksperimentalnoi i Teoreticheskoi Fiziki*, **87** (4), 1361–1370; [*Soviet Physics JETP*, **60** (4), 781–786].

118 Nikuni, T., Oshikawa, M., Oosawa, A., and Tanaka, H. (2000) Bose–Einstein condensation of dilute magnons in $TlCuCl_3$. *Physical Review Letters*, **84**, 5868–5871.

119 Radu, T., Wilhelm, H., Yushankhai, V., Kovrizhin, D., Coldea, R., Tylczynski, Z., Lühmann, T., and Steglich, F. (2005) Bose–Einstein condensation of magnons in Cs_2CuCl_4. *Physical Review Letters*, **95**, # 127 202, 4p.

120 Mills, D.L. (2007) Comment on "Bose–Einstein Condensation of Magnons in Cs_2CuCl_4" *Physical Review Letters*, **98**, 039-701.

121 Rakhimov, A., Mardonov, S., and Sherman E.Ya. (2011) Macroscopic properties of triplon Bose–Einstein condensates. *Annals of Physics*, **326**, 2499–2516.

122 Moskalenko, S.A. and Snoke, D.W. (2000) *Bose–Einstein Condensation of Excitons and Biexcitons*, Cambridge University Press, Cambridge.

123 Butov, L.V., Lai, C.W., Ivanov, A.L., Gossard, A.C., and Chemla, D.S. (2002) Towards Bose–Einstein condensation of excitons in potential traps. *Nature*, **417**, 47–52.

124 Borovik-Romanov, A.S., Bunkov, Yu.M., Dmitriev, V.V., and Mukharskiy, Yu.M. (1984) Long-lived induction decay signal investigations in ^{3}He. *Pisma v Zhurnal Eksperimentalnoi i Teoreticheskoi Fiziki*, **40** (6), 256–259; [JETP Letters, **40** (6), 1033–1037].

125 Fomin, I.A. (1984) Long-lived induction signal and spatially nonuniform spin precession in ^{3}He-*B*. *Pisma v Zhurnal Eksperimentalnoi i Teoreticheskoi Fiziki*, **40** (6), 260–262; [*JETP Letters*, **40** (6), 1037–1040].

126 Volovik, G. (2008) Twenty Years of Magnon Bose Condensation and Spin Current Superfluidity in ^{3}He-*B*. *Journal of Low Temperature Physics*, **153** (5), 266–284.

127 Bunkov, Yu.M. and Volovik, G.E. (2011) Spin Superfluidity and Magnon Bose–Einstein Condensation, arXiv:1003.4889v1 [cond-mat.other] (last accessed at: 25 March 2010).

128 Feldman, E.B. and Khitrin, A.K. (1994) The emergence of magnetic order in a nuclear spin system during adiabatic demagnetization. *Zhurnal Eksperimentalnoi i Teoreticheskoi Fiziki*, **106** (5), 1515–1524; [*Soviet Physics JETP*, **73** (5), 819–823].

129 Kalafati, Yu.D. and Safonov, V.L. (1989) Possibility of Bose condensation of magnons excited by incoherent pump. *Pisma v Zhurnal Eksperimentalnoi i Teoreticheskoi Fiziki*, **50** (3), 135–136; [*JETP Letters*, **50** (3), 149–151].

130 Kalafati, Yu.D. and Safonov, V.L. (1993) Theory of Bose condensation of magnons excited by noise. *Journal of Magnetism and Magnetic Materials*, **123**, 184–186.

131 Lavrinenko, A.V., Melkov, G.A., and Falkovich, G.E. Mutual influence of the kinetic and parametric methods of exciting spin waves. *Zhurnal Eksperimentalnoi i Teoreticheskoi Fiziki*, **87** (1), 205–211; [*Soviet Physics JETP*, **60** (1), 118–122].

132 Taranenko, A.Yu., Mino, M., Yamazaki, H., and Safonov, V.L. (1996) Electromagnetic emission by a system of nonequilibrium magnons in a ferrite. *Journal of the Physical Society of Japan*, **65** (12), 4072–4075.

133 Demidov, V.E., Dzyapko, O., Demokritov, S.O., Melkov, G.A., and Slavin, A.N. (2008) Observation of spontaneous coherence in Bose–Einstein condensate of magnons. *Physical Review Letters*, **100**, # 047 205 4p.

134 Chumak, A.V., Melkov, G.A., Demidov, V.E., Dzyapko, O., Safonov, V.L., and Demokritov, S.O. (2009) Bose–Einstein condensation of magnons under incoherent pumping. *Physical Review Letters*, **102**, # 187 205 4p.

135 Landau, L.D. and Lifshitz, E.M. (1975) *The Classical Theory of Fields*, Course of Theoretical Physics, Pergamon Press, Oxford.

136 Fröhlich, H. (1968) Long range coherence and energy storage in biological systems. *International Journal of Quantum Chemistry*, **2**, 641–649.

137 Fröhlich, H. (1977) Long range coherence in biological systems. *Rivista del Nuovo Cimento*, **7** (3), 399–418.

138 Hillebrands, B. and Ounadjela, K. (eds) (2003) Spin Dynamics in Confined Magnetic Structures II, Springer-Verlag, Berlin.

139 Sellmyer, D. and Skomski, R. (eds) (2006) Advanced Magnetic Nanostructures, Springer Science + Business Media Inc., New York.

140 Demokritov, S.O. (ed.) (2009) Spin Wave Confinement, Pan Stanford Publishing, Singapore.

141 Kruglyak, V.V., Demokritov, S.O., and Grundler, D. (2010) Magnonics. *Journal of Physics D: Applied Physics*, **43**, # 264 001 14p.

142 Suhl, H. (2007) *Relaxation Processes in Micromagnetics*, Oxford University Press, Oxford.

143 Silva, T.J., Kabos, P. and Pufall, M.R. (2002) Detection of coherent and incoherent spin dynamics during the magnetic switching process using vector-resolved nonlinear magneto-optics. *Applied Physics Letters*, **81**, 2205–2207.

144 Safonov, V.L. and Bertram, H.N. (1999) Magnetization reversal as a nonlinear multimode process. *Journal of Applied Physics*, **85**, 5072–5074.

145 Safonov, V.L. and Bertram, H.N. (2001) Spin-wave dynamic magnetization reversal in a quasi-single-domain magnetic grain. *Physical Review B*, **63**, # 094 419 11p.

146 Safonov, V.L. and Bertram, H.N. (2001) Magnetization dynamics and thermal fluctuations in fine grains and films, in *The Physics of Ultra-High-Density Magnetic Recording*, (eds M. Plumer, J. van Ek, and D. Weller), Springer-Verlag, Berlin, 81–109.

147 Boerner, E.D., Bertram, H.N., and Suhl, H. (2000) Dynamic relaxation in thin films. *Journal of Applied Physics*, **87**, 5389–5391.

148 Mao, C.Y., Zhu, J.G., and White, R.M. (2000) High speed characteristics of thin film write heads at deep submicron

track width. *Journal of Applied Physics*, **87**, 5416–5418.

149 Safonov, V.L. and Bertram, H.N. (2002) Nonuniform thermal magnetization noise in thin films: Application to GMR heads. *Journal of Applied Physics*, **91**, 7279–7281.

150 Dobin, A.Yu. and Victora, R.H. (2003) Intrinsic nonlinear ferromagnetic relaxation in thin metallic films. *Physical Review Letters*, **90**, # 167 203 4p.

151 Dobin, A.Yu. and Victora, R.H. (2004) Intrinsic nonlinear ferromagnetic relaxation. *Journal of Applied Physics*, **95**, 7139–7144.

152 Safonov, V.L. (2004) Microscopic mechanisms of magnetization reversal. *Journal of Applied Physics*, **95**, 7145–7150.

153 Krivosik, P. and Patton, C.E. (2010) Hamiltonian formulation of nonlinear spin-wave dynamics: Theory and applications. *Physical Review B*, **82**, # 184 428 27p.

154 Villain, J. (1974) Quantum theory of one- and two-dimensional ferro- and antiferromagnets with easy magnetization plane. I. Ideal 1-D or 2-D lattices without in-plane anisotropy. *Journal de Physique* **35**, 27–47.

155 Bar'yakhtar, V.G. and Yablonskii, D.A. (1975) On a representation for spin operators and its use in the theory of magnetism. *Teoreticheskaya i Matematicheskaya Fizika*, **25** (2), 250–259; [*Theoretical and Mathematical Physics*, **25** (2), 1109–1115].

156 Mansuripur, M. (1995) *The Physical Principles of Magneto-Optical Recording*, Cambridge University Press, Cambridge.

157 Fiorani, D. (ed.) (2005) Surface Effects in Magnetic Nanoparticles, Springer Science + Business Media, New York.

158 Gatteschi, D., Sessoli, R., and Villain, J. (2006) *Molecular Nanomagnets*, Oxford University Press, Oxford.

159 Gubin, S.P. (ed.) (2009) Magnetic Nanoparticles, Wiley-VCH, Weinheim.

160 Khitrin, A.K., Ermakov, V.L., and Fung, B.M. (2002) Information storage using a cluster of dipolar-coupled spins. *Chemical Physics Letters*, **360**, 161–166.

161 Khitrin, A.K., Ermakov, V.L., and Fung, B.M. (2002) Nuclear magnetic resonance molecular photography. *Journal of Chemical Physics*, **117**, 6903–6906.

162 Deya, K.K., Bhattacharyyaa,R., and Kumar, A. (2004) Use of spatial encoding in NMR photography. *Journal of Magnetic Resonance*, **171**, 359–363.

163 Khitrin, A.K., Ermakov, V.L., and Fung, B.M. (2002) NMR implementation of a parallel search algorithm. *Physical Review Letters*, **89**, # 277 902 4p.

164 Bhattacharyya, R., Ranabir Das, R., Ramanathan, K.V., and Kumar, A. (2005) Implementation of parallel search algorithms using spatial encoding by nuclear magnetic resonance. *Physical Review A*, **71**, # 052 313 6p.

165 Khitrin, A.K. and Ermakov, V.L. (2002) Spin Processor. arXiv:quant-ph/0205040v1 (last accessed at: 9 May 2002).

166 Provotorov, B.N. (1961) Magnetic resonance saturation in crystals. *Zhurnal Eksperimentalnoi i Teoreticheskoi Fiziki*, **41**, 1582–1591; [*Soviet Physics JETP*, **14**, 1126–1131].

167 Sokoloff, J.B. (1994) Theory of ferromagnetic resonance relaxation in very small solids. *Journal of Applied Physics*, **10** (10), 6075–6077.

168 Fernbach, S. and Proctor, W.G., (1955) Spin-echo memory device. *Journal of Applied Physics*, **26** (2), 170–181.

169 Anderson, A.G., Garwin, R.L., Hahn, E.L., Horton, G.W., Tucker, G.L., and Walker, R.M., (1955) Spin echo serial storage memory. *Journal of Applied Physics*, **26** (11), 1324–1338.

170 Gordon, J.P. and Bowers, K.D. (1958) Microwave spin echoes from donor electrons in silicon, *Physical Review Letters*, **1** (10), 368–370.

171 Govorkov, S.A. and Tulin, V.A. (1983) 2ω echo in a system with dynamic frequency shift (antiferromagnetic $MnCO_3$). *Pisma v Zhurnal Eksperimentalnoi i Teoreticheskoi Fiziki*, **37** (8), 383–386; [*JETP Letters*, **37** (8), 454–457].

172 Melkov, G.A., Kobljanskyj, Yu.V., Serga, A.A., Slavin, A.N., and Tiberkevich, V.S. (2001) Reversal of momentum relaxation, *Physical Review Letters*, **86**, 4918–4921.

173 Serga, A.A., Chumak, A.V., and Hillebrands, B. (2010) YIG magnonics, *Journal of Physics D: Applied Physics*, **43**, # 264 002 32p.

174 Khitrin, A.K. (2011) Selective excitation of homogeneous spectral lines. *The Journal of Chemical Physics*, **134**, 154 502 (7pp).

175 Khitrin, A.K. (2011) Long-lived NMR echoes in solids. *Journal of Magnetic Resonance*, **213** (1), 22-5.

176 Louisell, W.H. (1973) *Quantum Statistical Properties of Radiation*, John Wiley & Sons Inc., New York.

177 Walls, D.F. and Milburn, G.J. (1994) *Quantum Optics*, Springer-Verlag, Berlin.

178 Rezende, S.M. and Zagury, N. (1969) Coherent magnon states, *Physics Letters A*, **29** (1), 48.

179 De Araujo C.B. (1974) Quantum-statistical theory of the nonlinear excitation of magnons in parallel pumping experiments. *Physical Review B*, **10** (10), 3961–3968.

180 De Araujo C.B. (1976) On the coherence properties of parametric magnon states. *Physica Status Solidi (b)*, **75** (1), 327–332.

181 Wagner, M. (1986) Unitary Transformations in Solid State Physics, in *Modern Problems in Condensed Matter Sciences*, (eds. V.M. Agranovich and A.A. Maradudin), North-Holland, Amsterdam.

182 Safonov, V.L. (1982) On the theory of parametric excitation of waves in spin systems. (in Russian) *Preprint KIAE – 3691/1*, Kurchatov Institute of Atomic Energy, Moscow, 12p.

183 Safonov, V.L. (1983) Method of the spin hamiltonian diagonalization. *Physics Letters A*, **97** (4), 164–167.

184 Garanin, D.A. and Lutovinov, V.S. (1982) High-temperature spin wave dynamics of the uniaxial antiferromagnets. *Solid State Communications*, **44**, 1359–1362.

185 Andrienko, A.V., Ozhogin, V.I., Safonov, V.L., and Yakubovskii, A. Yu. (1983) Study of mechanisms of nuclear-spin-wave relaxation in the weakly anisotropic antiferromagnet $CsMnF_3$. *Zhurnal Eksperimentalnoi i Teoreticheskoi Fiziki*, **84** (3), 1158–1169; [*Soviet Physics JETP* **57** (3), 673–679].

186 Safonov, V.L. and Farzetdinova, R.M. (1991) On a possibility of magnon pumping in crystals with two-level defects. *Journal of Magnetism and Magnetic Materials* **98**, L235–L238.

187 Safonov, V.L. (1992) Theory of superconductivity for quasiparticles with parastatistics. *Physica Status Solidi (b)* **174**, 223–233.

188 Shi, Q., Safonov, V.L., Mino M., and Yamazaki, H. (1998) Unitary transformations in weakly nonideal Bose gases. *Physics Letters A*, **238**, 258–264.

189 Krasitskii, V.P. (1990) Canonical transformation in a theory of weakly nonlinear waves with a nondecay dispersion law *Zhurnal Eksperimentalnoi i Teoreticheskoi Fiziki*, **98** (5), 1644–1655; [*Soviet Physics JETP*, **71** (5), 921–927].

190 Krasitskii, V.P. (1994) *Journal of Fluid Mechanics* **272**, 1–20.

191 Bohm, D. (1959) General Theory of Collective Coordinates, in *The Many Body Problem*, Les Houches – session 1958, John Wiley & Sons, Inc., New York.

192 Wegner, F.J. (2001) Flow Equations for Hamiltonians. *Physics Reports*, **348**, 77–89.

193 Safonov, V.L. (2002) Continuous unitary transformations. arXiv:quant-ph/0202095v1 18 Feb 2002.

194 Safonov, V.L. and Bertram, H.N. (2003) Nonlinear microscopic relaxation of uniform magnetization precession. *Journal of Applied Physics*, **93**, 6912–6914.

195 Safonov, V.L. and Bertram, H.N. (2000) Impurity relaxation mechanism for dynamic magnetization reversal in a single domain grain. *Physical Review B*, **61**, R14 893–R14 896.

196 Safonov, V.L. and Bertram, H.N. (2005) Fluctuation-dissipation considerations and damping models for ferromagnetic materials. *Physical Review B*, **71**, # 224 402, 5p.

197 Callen, H.B. and Welton, T.A. (1951) Irreversibility and generalized noise, *Physical Review*, **83**, 34–40.

198 Perelomov, A. (1986) *Generalized Coherent States and Their Applications*, Springer, New York.

199 Mlodinov, L.D. and Papanicolaou, N. (1980) SO(2,1) Algebra and the Large N expansion in quantum mechanics. *Annals of Physics*, **128**, 314–334.

200 Gerry, C.C. and Grobe, R. (1997) *Quantum Semiclassical Optics*, **9**, 59–67.

201 Safonov, V.L. (2001) A single-mode realization of $SU(1,1)$. arXiv: quant-ph/0102126v1 23 Feb 2001.

202 Wiesenfeld, K. and McNamara, B. (1985) Period-doubling system as small-signal amplifiers. *Physical Revew Letters*, **55** (1), 13–16.

203 Wiesenfeld, K. and McNamara, B. (1986) Small-signal amplification in bifurcating dynamical systems. *Physical Review A*, **33** (1), 629–642.

204 Andrienko, A.V., Ozhogin, V.I., Podd'yakov, L.V., Safonov, V.L., and Yakubovskii, A.Yu. (1988) Amplification of small signals near the threshold of a collective-acoustic instability in a system of parametric nuclear magnons. *Zhurnal Eksperimentalnoi i Teoreticheskoi Fiziki*, **94** (1), 251–257; [*Soviet Physics JETP*, **67** (1), 141–144].

205 Cherepanov, V.B. (1986) Instability of parametrically excited spin waves interacting with acoustic waves. *Zhurnal Eksperimentalnoi i Teoreticheskoi Fiziki*, **90** (1), 153–157; [*Soviet Physics JETP*, **63** (1), 87–89].

206 Andrienko, A.V., Poddyakov, L.V., and Safonov, V.L. (1989) Study of threshold phenomena and collective effects in a system of parametric magnons and phonons by modulation of energy spectrum. (in Russian) *Sbornik nauchnykh trudov 1988*, Kurchatov Institute of Atomic Energy, Moscow, 58–59.

207 Andrienko, A.V. and Safonov, V.L. (1994) Observation of nonequilibrium Bose condensation of quasiphonons excited by a noisy microwave pump. *Pisma v Zhurnal Eksperimentalnoi i Teoreticheskoi Fiziki*, **60** (6), 446–451; [*JETP Letters* **60** (6), 464–469].

208 Andrienko, A.V. (2002) Observation of the coherent state in a system of nuclear spin-waves pairs excited by microwave noisy pumping. *Pisma v Zhurnal Eksperimentalnoi i Teoreticheskoi Fiziki*, **75** (2), 79–83; [*JETP Letters*, **75** (2), 71–75].

209 Andrienko, A.V. (2005) Noise pumping of nuclear spin waves in an antiferromagnet. *Zhurnal Eksperimentalnoi i Teoreticheskoi Fiziki*, **127** (1), 87–99; [*Journal of Experimental and Theoretical Physics*, **100** (1), 77–88].

Index

Nonequilibrium Magnons, First Edition. Vladimir L. Safonov.